THE WORK IN THE WORLD

THE WORK IN THE WORLD

Geographical Practice and the Written Word

MICHAEL R. CURRY

UNIVERSITY OF MINNESOTA PRESS
Minneapolis / London

Published by the University of Minnesota Press
111 Third Avenue South, Suite 290, Minneapolis, MN 55401-2520
Printed in the United States of America on acid-free paper

Library of Congress Cataloging-in-Publication Data
Curry, Michael R.
The work in the world : geographical practice and the written word / Michael R. Curry.
p. cm.
Includes bibliographical references (p. –) and index.
ISBN 0-8166-2664-2 (acid-free paper)
ISBN 0-8166-2665-0 (pbk. : acid-free paper)
1. Geography—Authorship. 2. Written communication. I. Title.
G70.C87 1996
808'.06691—dc20
96-16478

For Joanna

Contents

Preface

Like much else, academic life is filled with paradox, and anyone who has spent time in a university has seen it. There is the expert in technology who cannot program his VCR. There is the expert in exotic cultures who is incapable of making any headway in developing an understanding of her home department. There is the expert in ethics who bullies everyone in sight. And on and on; the reader can fill in other examples from his or her own experience.

My subject here is the nature of the written work in geography, and it, too, is paradoxical, although this fact is not often noted. Granted, one does find in academics a vague feeling of paradox in everyday discourse about the written work. There the aspiring academic learns early about the politics of publishing, about the regular interjection of political, subdisciplinary, and personal power into the process; at the same time, the student is taught to treat the written work with a kind of reverence, based on its ability to represent the world so clearly and transparently. And with a little nudging the average geographer might see an added paradox or two in the treatment of the geographical work. For one, since the 1980s we have seen a stream of works on this as text and that as text, on landscape or nature or city as text, but we have seen almost no works on "text as text." Further, while geographers have been hard at it, studying the geography of almost everything, no one has managed a geography of the written work, a geography of the text. Still, these paradoxical features of the written work have received little notice, and certainly no systematic treatment.

Here it would be easy enough to take the heroic stance: descrying what others have been too shallow to see, I have leaped into the breach. From an initial insight—there is a paradox here—the entire project has flowed forth. Like a detective I have searched first here, then there, for the answers, the source of the paradox. I have gathered facts, applied theories, reached conclusions—and here they are.

I am sure that someone, somewhere works in just this way. Alas, I do not. In fact, the extent of the paradoxes that surround the written work, and especially the written work in geography, only began to become clear to me late in the project. Why is this? Perhaps it is because I am a slow learner. This may be true, but it is surely more than simply a matter of saving face to say that there are other reasons for the difficulty in confronting this paradox. For like any true paradox, that of the nature of the written work is one that is self-supporting, even self-enriching. Indeed, and as I shall argue, it is the very *existence* of the written work that makes it so difficult to *understand* the written work.

Here the reader may be getting a feeling, a premonition of an argument looming in the background, ready to burst forth. This argument has, in the last dozen or so years, been common among an emerging group of theorists, call them postmodernists, late modernists, or perhaps hypermodernists. Among this crowd it has been fashionable to argue that we live in a world in which language fails us. And because all is lost, because all is futile, we may as well embrace this confusion and, yes, the ineluctability of paradox.

That I have managed five hundred words without resorting to neologism or to the casual use of parentheses and other punctuation marks ought to be some indication of my stand on this matter. As will become increasingly clear, I find this approach, this fey playfulness, to be both mistaken and troubling.

Why it is mistaken will become clear in the course of my argument. Why do I find it troubling? It is troubling because in its framing it places language so strongly in the foreground and thereby so values linguistic prowess, that it devalues and debases the real experience of paradox. For there can be no doubt that we live in a world deeply infected by paradox, a world in which by trying to get one thing we so often get something else entirely. The poverty-stricken Appalachian mother who impresses on her son the value of education sees him become more and more unlike her, more and more socially, culturally, and probably geographically distanced. The advocate of the destruction of totalitarian systems of political surveillance sees their decline, but with that decline the rise of market-based economies in which surveillance is far more pervasive, unregulated, and unpredictable. These are not paradoxes to be resolved by the use of a few parentheses and a couple of slashes. They are paradoxes that are far too fundamentally imbricated in everyday life, far too deeply embedded in the experience of things, to respond to such quick fixes.

Now, it may seem hyperbolic to suggest that what can be said of the relation of a parent to a child or a newly liberated person to an emerging society can be said of the relationship of a geographer to the written work. And it is true that the everyday experience of these works is dramatic in its ordinariness, that almost everything about such works seems routine, predictable, and decidedly unparadoxical. But the easy predictability of the written work, the naturalness of the ways in which we buy and sell, quote and cite, sits cheek by jowl with the quick anger with which we respond when a quotation is miswritten, a book borrowed and not returned. Indeed, what renders the written work so paradoxical and the paradox so important is just that it seems to support such divergent behaviors, and that it is so good at rendering invisible that divergence.

These paradoxes might be enough in themselves to make the written work a worthwhile topic of study. But more, whether in geography or elsewhere in science, the work itself stands at the border between professional practice and the world. True, the ritual awarding of prizes like the Nobel gives science a certain mien, just as the easy sonorities of celebrity scientists are its most visible popular manifestation. But by its very bulk the body of written works shows that scientists do what they are all expected to do—they produce objects, real and permanent. Forget theories, it is the book and the article that show that we are doing what we ought. And so, too, it is this same work that helps maintain the view that we can and do live in a world without paradox.

For my part, I do not particularly relish such a world, in which the written work has made it so easy for us to believe. But I do not react in this way because I believe those who write to be, somehow, doing a bad job. Rather, I react this way because the work itself conspires in complex ways to efface much that needs to be said and needs to be seen.

It may seem paradoxical, or ironic, or just incoherent and dishonest that I make these claims *about* the written work *in* a written work, and a conventional-looking one at that. Some would counter that I ought to take a different tack, write in a different way, use the punctuation to which I earlier referred, use eccentric typefaces, or write in neologisms. I ought in those ways to enlist the printed word as my ally in effecting a change in my reader. As will become clear, I find this by and large a misguided approach. Some, in contrast, might counsel me to avoid the issue entirely by claiming that "I wrote this for myself; if you do not understand it, too bad." No one who has spent the time and energy to write and publish a book can take this claim seriously.

Instead, I have taken a middle ground. As a consequence, it may at first glance appear that much of what I say is only obliquely related to the concerns of geography or of contemporary science. But I encourage the reader to see what follows here as a suggestion of a series of paths, which have led me to an appreciation of the richness and power of the written work, of its ability to represent and its ability to obscure. Most important has been that path that has led me to see the power of the work to deny just what one needs to understand it, an appreciation that the subject of inquiry needs to be the work in the world, and that what we need is in the most fundamental sense a geography of the written work.

Acknowledgments

This work was supported in part by grants from the Academic Senate of the University of California, Los Angeles. Portions of the research were carried out while I was in residence at the University of Edinburgh, and at the Center for the Critical Analysis of Contemporary Culture at Rutgers University.

During the project I have had research assistance from Karen Till, Terence Young, Camille Kirk, Barbara Cummings, and Travis Longcore; my thanks to all of them, and especially to Travis, who cleaned up the loose ends. Thanks also go to Joan Hackeling for editorial help.

Earlier versions of portions of this volume have been published elsewhere. I should like to thank the editors and publishers for permission to reprint, as follows: The discussions of authority in chapter 2 and of style and citation in chapter 5 are based upon "On the Possibility of Ethics in Geography: Writing, Citing, and the Construction of Intellectual Property," *Progress in Human Geography* 15 (1991): 125–47. The discussions of classification and ownership in chapter 2 are based upon "Shelf Length Zero: The Disappearance of the Geographical Text," in Ulf Strohmayer and G. Benko, eds., *Space and Social Theory: Geographical Interpretations of Postmodernity* (Cambridge: Blackwell, forthcoming). The discussion of postmodernism in chapter 3 is drawn from "Postmodernism, Language, and the Strains of Modernism," *Annals, Association of American Geographers* 81 (1991): 210–28. In chapter 4, I have based my discussion of Wittgenstein on "Forms of Life and Geographical Method," *Geographical Review* 79 (1989): 280–96, and "The Architectonic Impulse and the Reconceptualization of the Concrete in Contemporary Geography," in James Duncan and Trevor J. Barnes, eds., *Writing Worlds: Text, Metaphor, and Discourse* (New York: Routledge, 1991), pp. 97–117; the discussion of conceptions of space derives from "On Space and Spatial Practice in Contemporary Geography," in Carville Earle, Kent Mathewson, and Martin S. Kenzer, eds., *Concepts in Human Geography* (New York: Rowman and Littlefield, 1996), pp. 3–32. In chapter 6, the discussion of rights and responsibilities derives from "On the Possibility of Ethics in Geography: Writing, Citing, and the Construction of Intellectual Property," *Progress in Human Geography* 15 (1991): 125–47, and "Rights, Responsibilities, and Geographic Information Systems: Beyond the Power of the Image," *Cartography and Geographic Information Systems* 22 (1) (1995): 58–69.

In addition, earlier versions of these publications and of other unpublished material were presented at colloquia at University College London; the University of Technology, Loughborough; the University of Southern California; at the National Science Foundation Conference on the Social Impacts of Geographic Information, Friday Harbor, Washington; at the Conference on Geography among the Sciences at the University of Minnesota; and at the conference on Metaphor and Materiality: The Politics of Space and Nature, Rutgers University. My thanks go to those who offered comments on those presentations.

My understanding of the relationship between space and place has developed from discussions in my annual seminar on "The Historical Background of Modern Geography"; my thanks go to my students.

This work could not have been completed but for the continuing help and support of three longtime friends: Fred Lukermann, Mischa Penn, and Yi-Fu Tuan. Special thanks go to each; those who know them and their work will see the extent of my debt to them.

Nor could the work have been completed without the support of my friends and colleagues at UCLA; I should like especially to thank Nick Entrikin for help and encouragement well beyond the call of duty.

Finally, I owe a very different sort of thanks to my wife, Joanna, for her steady and enduring confidence.

Introduction

The last several years have seen an outpouring of works on the written work in science and in the academy more generally. There have been works on the rhetoric of science, on history as literature, on the sociology of the artifact, and on and on. Yet for a geographer two things are notable here. First, there has been little in the way of such works about the written work in geography. And second, there have been no *geographies* of the written work, whether geographical or otherwise.

Both of these facts should be of concern to the geographer. But although the first issue may be of little concern to those outside the discipline of geography, the second ought to be of some concern indeed. This is because the lack of a geography of the written work has been right at the heart of the difficulty that scholars have had in dealing with the thorniest of problems, ones that have bedeviled even the best and most sensitive of works on the written work. These problems are questions like, Why is it so difficult to change one's mind? Why is it so difficult to offer views that are truly radical? And why are works so often "misunderstood"? In this book I attempt to remedy that difficulty, to build on the work on the rhetoric and sociology of the written work and to show a way though this thicket of issues.

The immediate reaction here might very well be, How can a geography of the written work help? What can we really learn from a map of the distribution of books, or publishers, or authors, or readers? Will an economic model of the publishing profession really help? Should I be interested in the source of paper? The reader who hopes to find this kind of information in what follows will be disappointed. This is not at all the project that I have in mind, not at all what I mean when I say that what is needed is a geography of the written work. Rather, when I say that what is missing is a geography of the written work, I mean that what is missing is a study that attends seriously to those issues central to the discipline. What is missing is a work that explores the ways in which the written work is associated with particular places and with space, and what is certainly missing in those works that do exist is an appreciation of the difference between issues of place and issues of space.

Although there have been problems in coming to closure on the theoretical issues that surround the written work, I shall nonetheless argue that there is abroad a great deal of valuable knowledge about those works. Much of this is in the form of common sense. Over the course of an academic career, from school to university, in graduate school and beyond, the average person picks up a considerable store of information, stories and folklore, maxims and advice, about the written work. Although some of this can justly be derided as useless nonsense, much more cannot; indeed, it is just by accumulating this store of knowledge that one prepares to become an academic or a geographer, to write articles and books, to produce maps. All of us can with a little reflection come up with elements of this common sense, and most of us know people who were unable to acquire this knowledge, and whose careers have languished.

The existence of such common sense is surely not new; rather, it is as old as the written and then printed work, although *as* common sense it was not the sort of knowledge that was often put into print. As a consequence, when we look to earlier times, we have less material with which to work. Still, it seems fair to say that then as now there have been attempts to start from this common sense and from it to develop more structured and comprehensive accounts. But then or now, the move from common sense to theory has tended to take for granted a series of fundamentally geographical issues about the nature of place, the nature of space, and the interrelationship between the two.

This taking-for-granted has had a very specific form. It has been assumed that places can in the end always be seen as locations within a larger space, and it has been assumed that the move from space to place or from place to space is unproblematic in form. This may seem a minor point, one of relevance only within a very specific realm of questions, of which some are far removed from the question of the nature of the written work. But I shall argue quite the opposite. I shall argue that the adoption of this understanding of the relationship between space and place, the elision of the boundary between space and place, has great moment when it comes to the written work. And this is true for two reasons.

On the one hand, everyday actions are at once means of making places; just by engaging in everyday routines and habits we inevitably transform a bare world into a world of places. The making of places is a fundamental way in which we externalize our actions, in which we make those actions seem foreign, natural, and inevitable. And as a result, we render later actions easier here, more difficult there. Hence, the writing, reading, buying, selling, categorizing, or judging of a written work is always at the same time an act of making a place, or really of multiple places. An author makes a special sort of place, where labor creates an object whose conceit is that it can operate as a window on the author's ideas. But the work of the author is not all of a piece; there are subsidiary tasks, of quoting, citing, and editing, for example, and each of these is associated with a certain sort of place. The reader also creates a special sort of place, which operates with a parallel view; the place wherein a work is read is one in which the reader can by the act of reading engage "the mind" of an author far removed in space or even time. This engagement has traditionally been seen on the image *of* the image; if the author has succeeded, the work creates an image in the mind of the reader that is simply a reflection of an image projected by the author. A bookseller, too, creates a place, and it is not simply the shop where books are bought and sold. It is a larger place; to be a bookseller is to be party to the creation of the world as a place in which the very idea that there are books that are property to be sold, that contain ideas, that can be categorized makes sense. As important as these processes are, the confusion of space and place diminishes them, suggesting that we can best understand places not as the dynamic outgrowth of patterns of everyday actions but rather as concatenations of elements fixed in time. Still, once we begin to look more carefully at the ways in which people construct places, we shall begin to see the extent to which the elision of the boundary between space and place, by so often giving priority to space, renders invisible these very processes.

On the other hand, this elision has a related but different effect. If the view that one can move easily and unproblematically from the universal to the local and back obscures the fundamental connection between everyday practices and the places that they construct, it obscures matters in a second way. Where space is viewed as a grand and overarching structure, the written work emerges as in the first instance an object within that space, as do the author, the reader, and the rest. Here, too, is obscured the way in which the reference to space functions not as a conceptual "guarantor" but rather as an element in the practice of tying together diverse and disparate actions. Moreover, this appeal to a general space and a local place acts to undergird the belief that this image of structure and content, universal and particular, is more generally applicable. This in turn supports a set of beliefs about the nature of culture and mind, and about the act of following a rule and applying a theory. And these views, in turn, preordain a very particular, and only partial, understanding of the written work.

In what follows I shall address the nature of the written work in geography in these terms, by looking at what in geography is counted as the common sense about the written work. Here I take as uncontested the view developed by Clifford Geertz that common sense is a cultural construction (although at the same time I take issue with the notion that it constitutes a "system," just because to see it as a system is right at the outset to assume what needs to be explained).[1] If this common sense seems obvious, almost trivial, it is important to see that it is neither. What counts as the common sense about the written work is learned and might surely have been different, just as it may very well change. At the same time, this common sense does not simply arise from the world itself. Rather, it can better be seen as a matter of the appeal to simplifying images and stories, which themselves are borrowed from an available stock of images and stories about the way the world works.

In Part I, I shall look more closely at the way in which this store of images and ideas has been systematized into what I shall term the "traditional common sense." Until quite recently taken to be the last word on the nature of the written work, this view—which culminated in what we now call modernism—took as its starting point the very confusion of space and place, the very elevation of structure, that I mentioned earlier. First, I shall consider the development of the conventional view about the issue of representation, and especially the relationship between language and science. I shall trace its roots to Plato and Descartes and show the ways in which it is expressive of a deep ambivalence, which remains today. Here we shall find a powerful move toward the development of the view that the written work is a univocal container of ideas; we shall also see a set of counter moves, but moves that have been obscured by the compelling clarity of the dominant images.

Second, I shall turn to the ways in which the common sense about the written work came to be codified into a set of views taken to be descriptive of the place of the work in a broader social context. I shall show that the traditional view of the scientific work as a functional element within a larger structure derived from the same view of space and place that has been embodied in studies of representation. And I shall at the same time show the way in which changes like the invention of the citation index, the development of "big science," and the development of printing and then intellectual property regulation have acted to institutionalize, to crystallize this image of the place of the work in the world.

In Part II, I shall turn to a series of recent developments and to what is taken to be a major refiguration of the common sense about the written work. At once critical of the traditional common sense and productive of a new orthodoxy, this new common sense has quickly become codified into an encompassing structure.

There are three moments to this new common sense. First, language is rethought. The traditional notion that some language is literal and some figural, or figurative, is discarded; all is now seen as figural. The image of the text as a neutral and transparent representation of the world is abandoned; the written work is now seen as inevitably partial, obscuring just as it represents. Second, knowledge is now taken fundamentally to derive from a particular point of view, and the products of knowledge are thereby taken to be relative to that point of view. And third, the world is now seen as resistant to a reduction to a simple set of constituents. Rather, it is fundamentally messy; it consists of all manner of objects, events, and processes.

Together these changes in attitude toward meaning, knowledge, and that which exists have been at the heart of a rethinking of the entire project of science. Whereas in the traditional common sense it could be asserted that the scientific work could be trusted, as the author's genuine and open attempt to demonstrate the truth, to appeal to facts and reason, the work is now seen as fundamentally rhetorical, as an attempt to persuade. Indeed, many today would have it that the greater the appearance of neutrality, the more there is to suspect.

At the same time, for many it has become common sense that the work of science, and especially the conventional work of social science, is always an expression of superiority and domination, a declaration of the author's ability to know more about the subjects than the subjects know about themselves. If the guiding image in the traditional common sense was of the text as a map of the real, today many would see the image as very different; the written work is imagined as at once trying to seduce the reader, giving a come-hither look and treating the subject as a mere object, an other.

This new common sense has been developed in a number of disciplines. For one, there is a growing literature on the rhetoric of science, and now on the rhetoric of the social sciences. Beginning about fifteen years ago with a work in sociology by Joseph Gusfield, it burst forth in works on economics by Donald McCloskey and then in a long series announced by a collection edited by McCloskey, John S. Nelson, and Allan Megill.[2] Whether in science, the social sciences, or cartography, far from being the pure and pristine demonstrations that we had thought, relying on reason to do their task, the works of science have now been seen to be as much works of persuasion as those of any automobile salesman or politician.[3]

Similarly, a set of writers centered in anthropology have focused on the work as an intermediary in an often unseemly relationship between the author and the world. They have taken a critical attitude toward the author's relationship with the world, seeing the author as ineluctably a part of the world, rather than as someone observing it from a neutral vantage point. Beginning with Clifford Geertz, then continuing with James Clifford, George Marcus, and others, a large body of literature has developed; in this literature the text has become the villain, misrepresenting the power relationships inherent in the relationship between the author and the subjects of research, who in the process come to be treated as objects, as others.[4]

Indeed, although Geertz was critical, he has now been left in the wake by later work that has been much more polemical, arguing that the inequalities built into the relationship between the ethnographer and the subject need to be excised, and that a wholesale rethinking of the nature of the text is the first step. From the point of view of the anthropologist, the traditional ethnographic text now needs to be seen as intrinsically expressing power relations, with the author anthropologist in the structurally dominant position. Merely by writing such a text the anthropologist becomes a collaborator, an agent of a dominant Western society.

Now all of these views seem in a sense to involve a real recasting of our understanding of social scientific works. Whether in rhetoric, in anthropology, in philosophy, or in literary studies, the written work is being represented not as a neutral vehicle for "writing up findings," but rather as a device that at every turn is distorting, abusing, disempowering. We had thought that writing in science made things more clear; the opposite turns out to be true. Or so it is argued.

In chapter 3 I shall focus on this new common sense. And I shall show that although it provides a powerful antidote to the orthodoxy of the traditional common sense, it at the same time faces real difficulties. Perhaps most fundamentally, advocates of the new common sense have been unable to address adequately the very questions that I raised earlier, questions that the traditional common sense answered with facility, if with perhaps less success than might have been desired: Why is it so difficult to change one's mind? Why is it so difficult to offer views that are truly radical? And why are works so often "misunderstood"? I shall conclude that of the critics of the traditional common sense, feminist scholars have gone furthest in laying out an alternative that can address these questions.

In chapter 4 I shall begin to lay out some of the underpinnings of an alternative to the old and new common sense, an alternative that appears to me to be implicit in the ways in which some feminist scholars in geography have addressed questions about the discipline. Here I shall argue that in order better to understand the nature of the written work we need first to look separately at the concepts of space and of place; only by addressing these concepts separately can we begin to see the effects that their elision might have. With respect to space, we find that a series of very basic views have been remarkably robust in Western thought. Both in science and outside of it we find appeals to four different views. In one, space is rigid and hierarchical. In a second, it is abstract and absolute, as in the works of Newton. In a third, often associated with Leibniz, it is abstract but relational. And finally, it is common to see space not as something "out there," but rather as a construction of a knowing subject.

It is surely common to see certain of these views, and especially the Newtonian, as undergirding much of science. And indeed, when one speaks of the relation of space to place, it is common to imagine that places are simply locations within a big, empty Newtonian container. Moreover, it is common to imagine that ideas can best be seen on the same image, as objects in a conceptual space. And the written work is often conceived of as an object or commodity within an economic or classificatory space. But it is just these images, and the belief that one and only one can be in operation at a particular time, that make it so difficult to see more clearly the nature of the written work. For far from it being the case that historically one has replaced another, we find a persistence of each in different spheres of life, and each is supported by different social and technological arrangements.

The importance of this persistence will become more clear when we turn to a second issue, to the ways in which people construct places. Here we find that people create places in a wide variety of ways. They do so through official and unofficial rituals and routines. They do so by using language and symbols. And they do so by creating stories, narratives about those very places. It is commonly thought that the creation of a place is a matter of carving a niche out of some preexistent larger space, but this is not at all what happens; indeed, the issue of space arises most commonly when appeal is made to a particular notion of space as a unifying image, a means of putting together an otherwise disparate set of phenomena.

And for this reason, the elision of the difference between space and place here tends to draw attention in an unfortunate way away from the actions and routines that create places toward some larger "space" as the appropriate subject of discourse. This is especially unfortunate, as I have earlier pointed out, because the creation of places is a fundamental way in which people externalize their actions and establish environments that are seen as natural, necessary, and constricting. It is by creating a place within which the book may be read that one creates a set of limits on the nature of the book, and the same is true for the matter of authoring.

At the same time, there is a second effect of the uncritical adoption of the usual view of space and place. This view, because it seems to parallel the views of ideas as elements within a mental or logical space and of the application of theories and rules as a matter characterizable in terms of this same image, renders it difficult indeed to make sense of the nature of everyday practice. It suggests that at each juncture one needs to be able to step back and take a more abstracted view. And this, in turn, renders it all the more difficult for a student of the written work to take a reasoned view of the practices and objects with which one as an author and reader is necessarily engaged. It denies the possibility of reflexivity.

In addressing this issue I shall begin with the works of Wittgenstein and suggest ways in which one may develop a more adequate understanding of how actions and practices lead to what are typified as larger-scale structures. His work is a useful place to start for several reasons. For one, a look at students of human practices as diverse as Pierre Bourdieu, Michel de Certeau, Anthony Giddens, and Peter Winch shows that each begins with an engagement with Wittgenstein's work and with the questions that that work raises.[5] Each takes his work in a very different direction, but what is important is the agreement that an adequate understanding of the relationship between language and practice or action is fundamental to any study of human practices more generally and hence to the understanding of the ways in which the elaboration and sedimentation of those practices lead to the establishment of human institutions. At the same time, Wittgenstein's work stands out as offering a sustained analysis of the many ways in which the simplifying image and theory come to be seen as containing the rules for their own application, and thereby come to be seen as capable of maintaining an existence separate from the actions and institutions that comprise them. In this way, his work at the same time presents a powerful case for the essential inadequacy of attempts to account for human action in theoretical terms. Although Wittgenstein did not himself develop anything like a social science, his work offers a glimpse of an alternative, where the task of the social scientist is to approach the world with a repertory of questions and a critical eye and to attempt to discern the ways in which humans have structured the world by descrying the ways in which we develop patterns of practices and in which we come to see those patterns as independent and self-contained.

Finally, in Part III, I shall develop the approach that I have sketched out through a series of case studies. In chapter 5, I shall characterize the ways in which authorship is constructed within geography today. According to the traditional common sense, to be an author was to operate within a series of systems, notably the systems of representation, authority, and ownership. Especially in the system of representation, one had great freedom to move; the limits of one's ability to engage in scientific representation were defined in a fundamental sense as the limits of one's intellectual ability. And because the system of science was believed to be one that rewarded quality, at least when it was unencumbered by external influences, there was a mapping between the other systems and the system of representation, so that work that better represented the truth was better rewarded, and vice versa. By contrast, in the new common sense, science in the worst case—which is imagined to be all too often the case—is represented as a process of invention, one directed by the desire to accumulate what is sometimes called, to use Bourdieu's term, "intellectual capital."

But as I have argued, what these two views have in common is an inability to answer certain very fundamental questions. The traditional common sense seems to make it all too easy to change one's mind: one just does it. And having done so, having changed the theories or stories that one tells about the world, one becomes a different sort of geographer. The positivist becomes a Marxist or a humanist. Remarkably, we find much the same view when we turn to the new common sense; there anthropologists like Stephen Tyler and Vincent Crapanzano have made a similar argument, that one can and ought to alter one's relationship with the subjects of one's work by writing ethnography in a different way.[6]

And yet it seems clear that the converted positivist shares a great deal with her former self, and the author of experimental ethnographies with his. And I shall argue that to understand this we need to look at the ways in which these authors have constructed in their authoring a series of places. This construction occurs first through the use of a series of ways of writing, a series of conventions that mark one's work as "belonging" in a particular set of places, in journals and publishing houses.

The construction of places occurs in a second and related way as one represents oneself as having a certain relationship with one's work. Authors commonly represent their work simply as catalogs of facts, as reflections of the contents of their own minds, or as personal achievements, the result of personal journeys. Most works, in fact, embody elements of all three. Each of these reflects a particular notion of authoring, which involves the collection of facts, the establishment of a certain mental state, or the overcoming of a resistant world, and each notion in turn is associated with a set of practices intimately connected with a particular sort of place. Moreover, each is connected with a different set of justifications for the attribution of the rights of ownership to a work, and in this way each insinuates itself into deeply ingrained sets of institutions. Indeed, when we look at the written work from this perspective, we find that where we appear to find great changes, as in the development of the work of David Harvey, we actually find great continuity; by contrast, in the case of computer cartography and geographic information systems it might appear that there is great continuity with the past, where in reality the opposite is true.

Finally, any author through the choice of intellectual stance and method creates the possibility of alliances with others. Through an analysis of a series of programmatic papers, I shall show that the widespread success of one group, advocates of the quantitative revolution, can be seen as a result of their having advocated a view that completely restructured the places within which geographical research could be carried out. In doing so they opened the door for alliances with a great many previously unrelated individuals, technologies, and institutions, and these alliances allowed the quantitative revolution to succeed well after its intellectual vigor appeared spent.

In the final chapter, I shall direct my attention not to many works but to a single one, to Henri Lefebvre's *The Production of Space.* Widely praised, this work argued for a rethinking of the concept of space, a rethinking different from the one advocated here. Yet as self-conscious as Lefebvre attempted to be, on closer analysis we find that his work is shot through with a wide variety of spatial notions, notions with which he may have been quite uncomfortable.

The point here is not that Lefebvre was unintelligent. Rather, it is a deeper point, that Lefebvre rightly appealed to a variety of spatial practices and conceptions that he knew—at least implicitly—that his reader understood and expected, and that are part and parcel of the system of written works. Not to write within the context of those conceptions would be to risk not being understood. And so, just as Lefebvre's work is of interest because of its analysis of the nature of space, it is also of interest because of the way in which he assumes that his reader understands and operates within these spatial practices and conceptions. Although he appears to believe that he is writing for any reader, he is, in fact, writing for a reader who is very closely defined.

In the end, what I shall have offered is merely a set of questions to be asked and a commentary on a series of things that occurred in the past and seem likely to recur. That I have offered no more is not a result of some lack of will but rather is a direct consequence of what I have said about the written work. For if what I have said is true, it will turn out that *any* attempt to understand the nature of the written work through the application of the sort of theory that has been developed in the formalizations of the traditional common sense and the new common sense will be only partial, incomplete.

This incompleteness arises because the written work does not tell us only about the theoretical commitments that people explicitly make. Nor does it simply tell us about the ways in which they attempt to persuade. And nor does it simply tell us of the power relations that exist between author and subject. Rather, the written work can better be seen as a kind of fulcrum, in and through which the practices associated with the written work express that which the author finds possible, probable, certain—and beyond the pale. All works, geographical or not, do more: they tell us what the author believes about central issues within our discipline, about space and place and nature. Moreover, they tell us about the places of different individuals and groups in a way far richer than we otherwise find. If we look in the right way at the written work, we shall find that people who style themselves as diametrically opposed one to another turn out to be very much alike, just as we shall find that groups that see themselves as solidly homogeneous turn out to be deeply fractured.

And so, an Allen Pred or a Gunnar Olsson may bemoan the difficulty of communicating, but we shall find that in a range of ways they assume that communicating could be no simpler. An Henri Lefebvre may celebrate his grasp of the nuances of space, while at every turn appealing to ones that he fails to note. The developer of a geographic information system may see the systems as powerful continuations of the cartographic tradition, whereas what is being developed undercuts that tradition in unnoticed ways. And any geographer may proclaim a dramatic change of heart, a radical new way of looking at things, but one that turns out to be all too familiar.

Here one may wish to interject, but what of *this* project? Does it not fall victim to its own method? Have I exempted myself from the analysis that I make? And indeed, this is the crux of the matter. This is *precisely why* it will not do simply to construct and apply a theory. It will not do because the construction and application of a theory are not the isolated mental acts that a Descartes would have us believe them to be. To construct a theory, to write and publish an article or book, is at once to implicate oneself in a wide range of practices, to take as real a wide range of objects, to aver a set of conceptions of space and place, causality and the like, that simply cannot be avoided.

To suggest that "a theory" can capture those conceptions is to be misled. It is to be misled because the written works, and the practices that are associated with them, are not the products of a single mind, a single center of rational deliberation. What I write is never simply the outcome of a set of thoughts, conscious or not, emanating from my mind.

Rather, the written work is always the outcome of long processes of sedimentation, processes that can occur singly, in concert, and in contradiction. And to suggest that some theory will be able to grasp these processes and their wakes is to gravely underestimate the extent to which a single person can harbor what to the theoretician can only seem like pathological inconsistencies.

So to understand the written work in geography, as elsewhere, it is necessary that we see the work in the world. It is necessary that we attend to the practices and objects and institutions and images that make it what it is. And this is the point from which to address the question that I have raised: Does this project not fall victim to its own method? The answer is simple: it surely does. For I shall in what follows lay out a series of questions to ask in looking at a work, a series of tacks to take. I assume that the reader will apply what I have said to what is being read.

But in not *saying* what my biases are, have I not kept something important hidden? Have I not exempted myself from the analysis that I make? If what I am arguing is true, I have not. For to suggest that it takes some verbal act to get across to the reader the "essential" features of a work is to miss the point quite entirely. What we think about space and nature is not locked up in our minds, not encased in some Cartesian homunculus. What we think is there to be seen, in the ways we write, argue, cite, publish, and on and on. And what obscures this is not some philosophical intractability, not the fact that the mind is some invisible substance, there only to be inferred. What obscures it, instead, is the ease with which we turn from the complexity of the practices with which the writer engages the world and look for the easy image, the essence, the one story that will turn what was rough and confusing into something tamed.

PART I

The Traditional Common Sense: The Universal Text in a Universal System

CHAPTER ONE

Formalizing Common Sense (I): The Text as Representation

There is a commonsense way of looking at the written work, which sees the work as in the first instance devoted to the task of representing something about the world. Because representation is the central concern, the analysis of the written work focuses on the interrelationships among the work, its author, its author's ideas, the reader, the reader's ideas, and the world itself. This list may seem long, but in fact the image of the representing function of the written work is often quite simple. An author/researcher observes the world and forms a set of mental ideas. The mental ideas are transferred into language and transferred to a printed page. The reader reads the page and thereby appropriates the author's ideas; with the author's ideas "in mind," the reader can now see the world as the author did.

If this process also seems complex, in fact it is often represented more simply in visual form, as a series of mappings, of world to author to written work to reader and back to the world. Indeed, a great deal has recently been made of this image of knowing as a kind of mapping or mirroring; certainly one of the most widely cited recent works of philosophy has been Richard Rorty's *Philosophy and the Mirror of Nature,* which explicitly decried this view.[1] And Rorty and others have been right at the heart of a movement that I shall describe as having formulated a new common sense, and that saw this view as both utterly dominant and variously pernicious.

But if we look in more detail at the development of this view, at the codification of the traditional common sense, we shall find a more complex situation than might first appear. Certainly the view that the mind and the written work mirror or map onto nature did not appear full blown; in fact, it developed over a period of time and through a series of stages. These individual stages are seldom univocal. We shall find in the development of the understanding of the written work that at each stage the very individuals who are most closely associated with a particular stage have at the same time expressed views quite counter to the images that we typically have of them.

These stages need to be seen against the background of wider social and technological changes, and this ought to suggest something more; the development of one stage is seldom a matter of its replacing another but is more usually a matter of a new way of thinking emerging alongside an older one. Why have these contrary undercurrents remained submerged? It seems to me that they have been submerged because they run counter to a dominant and powerful set of images that were supported in a variety of ways. So as we look at the development and formalization of the commonsense way of looking at the written work, we find some elements of everyday lore and wisdom being moved to the background, others moved to the foreground. And in each case the elements that have been moved to the foreground are those that sit most easily with a universalizing view of space and a view that confuses place with space. The result has been the acceptance of a particular view of the written work as a nonlinguistic object, and perhaps more important, the implicit appeal to a "universal text," one that subsumes all individual texts.

This acceptance is a consequence of three sets of developments, each manifest in broader currents in Western thought. One was the development of writing and of the split between written and oral cultures. The second was the development of the view that in the (achievable) ideal case, written works do not appeal to particular audiences but rather are written for and understandable by all users of a particular language. Here it is seen as possible to decontextualize language, to think of texts as having a kind of autonomous existence. And the third was the development of the notion that knowledge, again in an achievable ideal case, consists of an ordered system of concepts or ideas that exist in a system that is single and simultaneous and hence can be grasped all at once.

In each of these three cases, we see the development of a view of the written work that involves a particular notion of the relationship of place to space. For Plato the key to knowledge was dialogue, dialogue always in a particular place; for him the academy was the place where one was able to engage in dialectic and gain knowledge. At the same time, in Plato we find an ambivalence. If dialogue in places was important to him, he is better remembered for the development of a form of realism where there is assumed to exist a permanent and transcendent world of ideas; this realism, closely tied to his idealization of mathematics, has been a fundamental source of the view that we can and should turn away from the particular. And this view has obscured the notion of the gaining of knowledge as a process that fundamentally occurs in a place.

In Descartes there is also ambivalence, a view that language may be either central or irrelevant. But this ambivalence, even more than in Plato, is obscured by a central and powerful image, of the mind as existing in a kind of universal space. After Descartes one might imagine science as, in principle, a universal text, but the individual work was reduced to a point.

In this chapter I shall lay out the nature of these three developments, each of which is central to what I have termed the *image of representation.* In doing so I shall pay special attention to the work of Descartes; in the matter of texts his work forms an important turning point, and he, in fact, offers two conceptions of the relationship between language and science, both of which—although they are conflicting—have been crucial to the understanding of representation. In the next chapter we shall see that this same image has reverberated into discussions of the social structure of science itself.

Words Frozen in Time: The Legacy of the Platonic Dialogues

The first development is the creation of a split between oral and written cultures. In the past few years much has been written about this split; Jack Goody, Eric Havelock, and Walter Ong have made it the focus of their work.[2] All three have argued that the development first of writing and then of printing has led to a massive reorientation of cultures, one in which the very ways in which people think have been changed. As Ong so memorably put it, with the development of written culture "the whole intellectual world goes hollow."[3]

According to Havelock, this change was already in process at the time of Plato. In earlier oral cultures knowledge had been evanescent; its preservation required constant attention, and as Milman Parry showed in his study of Homer, the use of repetition and cliché as mnemonic devices was a favored way of that preservation.[4] In contrast, by the time of Plato writing was an entrenched part of culture, and Plato's own work is expressive of this change. As a result, according to Havelock, it was no longer necessary to attend constantly to the maintenance of traditional knowledge; as a result, knowledge could be maintained in an external, physical form, and it became possible to turn attention to analysis and criticism.

As Havelock and others have noted, Plato was ambivalent about this development. On the one hand, he *wrote* works and intended that they be *read*; on the other—in the *Phaedrus*,[5] for example—he has Socrates criticizing writing in the following way:

> And so it is that you, by reason of your tender regard for the writing that is your offspring, have declared the very opposite of its true effect. If men learn this, it will implant forgetfulness in their souls: they will cease to exercise memory because they rely on that which is written, calling things to remembrance no longer from within themselves, but by means of external marks. (*Phaedrus*, 274e–275a)

It might appear here that Plato is merely being an early Luddite, making the same sort of argument that we see today in the matter of electronic calculators and computers. But his argument is more complex:

> And it is no true wisdom that you offer your disciples, but only its semblance; for by telling them of many things without teaching them you will make them seem to know much, while for the most part they know nothing; and as men filled, not with wisdom, but with the conceit of wisdom, they will be a burden to their fellows. (*Phaedrus*, 275a–275c)

Reading cannot lead to knowledge, but only to something that looks like it, to the "true belief" that he so criticized:

> Then anyone who leaves behind him a written manual, and likewise anyone who takes it over from him, on the supposition that such writing will provide something reliable and permanent, must be exceedingly simple minded. (*Phaedrus*, 275c–d)

> You know, phaeton, that's the strange thing about writing, which makes it truly analogous to painting. The painter's products stand before us as though they were alive: but if you question them, they maintain a most majestic silence. It is the same with written words: they seem to talk to you as though they were intelligent, but if you ask them anything about what they say, from a desire to be instructed, they go on telling you the same thing forever. (*Phaedrus*, 275d)

Written works lack the central feature of that which can lead to knowledge; they cannot engage in dialogue, but rather they respond to criticism simply by repeating themselves. Here they are in stark contrast to another sort of discourse, "the sort that goes together with knowledge, and is written in the soul of the learner: that can defend itself, and knows to whom it should speak and to whom it should say nothing. [It is] living speech, the original of which the written discourse may fairly be called a kind of image" (*Phaedrus*, 276a).

It is important, though, to see that only certain forms of "living speech" are acceptable, whereas others are just as bad as the written. Singled out for criticism is rhetoric. Indeed, so devastating is Plato's critique of rhetoric that some modern-day rhetoricians go so far as to blame him for the long eclipse of their discipline. Brian Vickers, for example, devotes a long chapter to Plato; accusing Plato of using tactics to "outwit" rather than convince, he lauds a recent critical study of the *Gorgias* as a "dense mass of commentary" containing "recurring criticisms" that are "devastating."[6] Leaving no argument unanswered, he claims that the distinction made by Plato between the mind and the body is "less neutral classification than polemical discrimination."[7]

Works like those of Vickers—and he is not alone—ought not to blind us to the seriousness of the long-standing critique offered by Plato, which would appear to have had a significant impact on the possibility of applying the tools of the rhetorician to the study of scientific texts. The groundwork of this critique is the distinction, central throughout his work, between the worlds of Being and Becoming and of reality and appearance, and between the parallel processes of knowledge and belief. In order truly to have knowledge it is necessary that one look beyond the everyday, changing world—of which it is possible only to have beliefs—and grasp the nature of that which informs it and makes it what it is, the underlying world of that which does not change, of reality. It is only of the unchanging that we can truly have knowledge.

Here Plato contrasts "the Sophist [who] takes refuge in the darkness of Not-being, where he is at home and has the knack of feeling his way"[8] with the dialectician, who "divide[s] according to Kinds, not taking the same Form for a different one or a different one for the same" (*Sophist*, 253d), and who can "distinguish, Kind by Kind, in what ways the several Kinds can or cannot combine" (*Sophist*, 253e).

It is important to see that a "knack," for Plato, is a set of habits or rules, but ones that are fundamentally suspect because the aim of a knack is the satisfaction of a goal other than the grasping of the nature of the world of being. In the case at hand, the practice of oratory and the application of rules of rhetoric is a matter of just such a knack,[9] which is all the worse when the "beauty and diversity of their words bewitches our souls."[10] Indeed, this is the very aim of rhetoric—to "influence men's souls" (*Phaedrus*, 271c). In fact, there is a sense in which an orator must be rather like a social scientist:

> [T]he intending orator must know what types of souls there are. Now these are of a determinate number, and their variety results in a variety of individuals. To the types of soul thus discriminated there corresponds a determinate number of types of discourse. Hence a certain type of hearer will be easy to persuade by a certain type of speech to take such-and-such action for such-and-such reason, while another type will be hard to persuade. And when he is competent to say what type of man is susceptible to what type of discourse; when, further, he can, on catching sight of so-and-so, tell himself "That is the man . . . and in order to persuade him of so-and-so I have to apply *these* arguments in *this* fashion" . . . then and not till then has he well and truly achieved the art. (*Phaedrus*, 271d–272a)

If rhetoric, if argument aimed at persuading an audience, does not lead to knowledge, to what does it lead? It leads, he argues, only to belief:

> *Socrates.* Now, rhetoric produces persuasion in lawcourts and other public assemblies about things which are just and unjust. What kind of persuasion? The kind from which arises belief without knowledge, or the kind from which arises knowledge?
> *Gorgias.* Clearly, Socrates, the kind from which arises belief. . . .
> *Socrates.* Therefore, the rhetorician . . . only creates belief. For after all, it would be impossible to instruct such a large crowd in a short time about matters of such importance. (*Gorgias*, 454e–455a)

It becomes clear that the issue is not simply one of certainty, where knowledge might be seen as certain or even true belief. Indeed, Plato presents what he takes to be a radically different notion of knowledge, where the acquisition of knowledge is a matter of self transformation:

> The way each of us learns compares with what happens to the eye: it cannot be turned away from darkness to face the light without turning the whole body. So it is with our capacity to know; together with the entire soul one must turn away from the world of transient things toward the world of perpetual being, until finally one learns to endure the sight of its most radiant manifestation.[11]

This transformation is not a matter of the memorization of true facts; rather it requires the development of the ability to engage in dialectic. And this development requires dialogue, which is difficult, almost impossible, to carry out in a group: "I dismiss a multitude of witnesses because I can't carry on a discussion with a crowd, but I know how to put the vote to one" (*Gorgias*, 474a–b).

In the matters of intellectual transformation and of dialogue the place of the written work comes to the fore. But it does so in a way that for contemporary readers may appear confusing. Time and again Plato—in the guise of Socrates—asks about the nature of things, of the real world, or of justice or happiness. And it seems from the ways in which he asks his questions, as it does from the weight of evidence of his written work, that behind his questions lie a series of views about what the answers are to be. In some cases more than others it appears that he has genuine perplexities, but overall it appears that he has what we, today, might want to term a *world view*—although he certainly would object to the term. Now, his possession of a world view would seem to suggest that we can address his texts with questions like, What *is* the nature of justice? and What *is* the nature of the real world? And it would suggest that there is an answer to be had.

But this is just what he wishes to deny by choosing the form of the texts, which are dialogues. The written works are merely the occasions for the development of knowledge, and not the sources of that knowledge. For Plato, the function of the dialogues is to force the reader to gradually hone a set of mental tools that will come to be applied to these issues. In doing so the reader strips away beliefs and comes gradually to see the world "as it is." Hence, although this transformation may occur through a process of dialectic addressed *to* texts, it is not a result simply of a reading *of* those texts.

In fact, for Plato the texts are not representations of reality, although they do include statements about reality. Although we may think of a work of representation as one in which the text expresses the author's ideas and thus makes publicly available to the audience the author's image of the world, texts do not do that. Moreover, they do not even delineate a picture of the process of knowing. Rather, they offer the reader two different—and conflicting—things.

First—and this was clearly Plato's primary intention—they offer an alternative image of the community of knowers and the means to its access. In an important sense Plato, as Havelock and Ong have suggested, was offering an alternative to traditional, oral ways of knowing and of creating community. In traditional or preliterate societies the means of storage is memory, the means of transmission oral. The aid to storage is very different; it is the image of place. In rhetoric the types of arguments available to the rhetor are stored, in memory, in places (*topoi*), and according to Cicero's recitation of the story of Simonides, memory and place are inextricably linked. Transmission is through the use of certain schemes of meter and rhyme and repetition that allow a general theme to be replayed with variations over a long period, and to be easily transmitted from one individual or group to another. In cases where oral transmission is the primary means, the content is typically a story or narrative, one in which time is the medium. Explanations of social and natural events are fundamentally narrative, and it is difficult—Ong would say impossible—to transmit sustained abstract arguments.

In redefining knowledge, though, Plato's texts reject both memory and oral transmission—and they offer an alternative. The alternative to oral transmission of knowledge is a method for constantly recovering it, and that method is dialectic. A person who has somehow forgotten the nature of justice will be able, by adopting the correct state of mind, to come to a conclusion. Indeed, the same is the case in the matter of mathematics, as Plato demonstrates in the *Meno,* where a "common slave boy" is led by the process of dialectic to come to a realization of the understanding of a sophisticated geometrical proof.[12]

Much the same is true in the matter of memory. Centered around his assertion of the immortality of the soul, in Plato recollection takes over what is in rhetoric the central place of memory. Here the fundamental issue is the way in which one can, by properly training the "mind's eye," see eternal truths, both about justice and the good and about the structure of the world. If the art of memory proceeds through the association of certain memories with certain places, recollection is simply a matter of clearing away the clutter of the everyday world and seeing clearly what must be. Although in one sense, then, Plato is offering an alternative to traditional oral culture, he is in fact offering in his written works an image of a way to maintain through oral means a defense against texts becoming as hardened as the oral traditions that he hoped they would replace.

Notwithstanding this, the notion that he has offered of the mind's eye is at the fountainhead of a tradition of visual metaphors for knowing, which in Descartes come to be solidified into what Heidegger has called the possibility of seeing the world as a picture. In an important sense it is this latter image of the mind's eye that is the legacy of Plato; it is this to which we point when we see him as operating at the beginning of the Western scientific tradition. And this visual tradition in a variety of ways devalues the written work, whether obliquely in images of the mind's eye, of memory, and of recollection, or directly, in the *Cratylus,* where Plato dismisses the possibility of there being a language in which "a name is an instrument of teaching and of distinguishing natures, as the shuttle of distinguishing the threads of the web,"[13] and concludes that "we must rest content with the admission that the knowledge of things is not to be derived from names.

No; they must rather be studied and investigated in their connexion with one another."[14]

Unfortunately, this view of Plato's relation to—and rejection of—the written work has blinded most authors to those ways in which Plato makes absolutely central to knowledge a particular form of text. At best the result has been to see this form only in philosophy, where the dialogue now and again is revived, and where writers like Nietzsche and Wittgenstein have aphoristically attempted the same, or in geography, where only Gunnar Olsson, Dagmar Reichart, and perhaps Allan Pred have engaged in the creation of this sort of text.[15] But as I shall argue, the dialogical element of the text lives on, and particularly in the conversations that lie implicit in texts. I mean here not so much the article—commentary—reply form that we often see, but rather the more simple (and ubiquitous) conversations carried out with others in citations and references, and with oneself in footnotes. I shall return to these conversations in later chapters.

The Double Legacy of Descartes

In *The Order of Things*, Foucault describes the development of modernism and points to Descartes as its founder.[16] It is Descartes, he argues, whose views of the subject and of knowledge so permeate Western thought until the fall of classicism at the end of the eighteenth century. And Foucault is not alone in this view; in his essay "The Age of the World Picture," Heidegger argues that it is in Descartes that we first find the possibility of viewing the world itself as a picture, as something that can be encompassed in a single image, and in the image of a system.[17]

In both of these works, though, one finds a problem similar to that which we found in Plato: their readings portray Descartes in a way that at once assumes a particular notion of the written work and occludes the ambiguity within those texts themselves. To put matters simply, Foucault and Heidegger give Descartes a modernist or classicist reading that his work does not entirely deserve. His works become models of the modernist text, but this is only possible, as we shall see, by applying to them modernist readings. In these modernist readings much that is interesting is lost. And a part of what is lost is the possibility in the sciences of taking the written work seriously.

There are, of course, those who have attempted to portray Descartes as a more complex character. Stephen Toulmin, for example, has argued that Descartes needs, at the very least, to be seen as standing at the end of the creation of modernism as an intellectual tradition, rather than as modernism's inventor.[18] He argues that in Descartes, as opposed, for example, to Montaigne's essays of the 1570s and 1580s, we find a kind of ossification, an attempt to build an intellectual scheme that will shore up what looked to be a crumbling political and social world. Far from opening up a field of inquiry, Descartes encloses that field within the strictures of a rigid system. In this, of course, there are echoes of Foucault and Heidegger, alongside the tempering of the Foucauldian notion of rupture.

And more to the point here, Dalia Judovitz has argued that there are deeper problems with the usual picture of Descartes.[19] Indeed, she argues, we can find in his work powerful strains of the Baroque. In particular, she directs our attention to two areas. First, on the one hand Descartes appears to represent a new way of thinking about the mind, but on the other we find deep strains of traditional ways of thinking about memory and rhetoric. And second, although Descartes is pointing to the development of the notion that we can see the world as a picture, this notion is not entirely new; we find much the same in the Baroque. What *is* new is that unlike the Baroque picture of the world, Descartes's is not visual or conventional; rather, it is schematic and universal.

Both Toulmin and Judovitz have noted qualifications of the modernist reading of Descartes. Yet if they have pointed to important features of the social and historical contexts within which he was writing, they have not enunciated the complex role of the written work in Descartes, a role that involves sometimes contradictory notions of language, ideas, memory, simultaneity, system—and the subject. In fact, these notions become intertwined in not one, but two notions of the text. The first arises from an understanding of the possibility of a perfect or universal language; this notion persists today in chemistry in the periodic table, in the use of equations, and in the usual understanding of the "language of maps." The second notion of text is in a sense quite the opposite; it arises from Descartes's denigration of language. He proposes what I call the nonlinguistic text, a written work that exists only as a part of a larger system of ideas, subjects, and the world. This is the view to which Foucault and Heidegger have pointed, and it is the more often noticed of the two. But to understand the complex legacy of modernism, we need to attend to both.

THE UNIVERSAL LANGUAGE AND THE UNIVERSAL TEXT

One way in which Descartes's legacy renders the study of the written work irrelevant is by setting up as an ideal a perfect text, one in which the warts and blemishes of all actual texts have been removed, rendered ineffectual. When in 1628 he wrote the *Rules for the Direction of the Mind,* Descartes was writing against the background of a pervasive view in which language was seen as imperfect. From one point of view, a Christian one, this was a result of the fall of man. For Adam language was truly communicative; God brought all of the animals before him "to see what he would call them: and whatsoever Adam called every living creature was the name thereof..." (Genesis 2:19). "The whole earth was of one language and one speech" (Genesis 11:1). But after the Tower of Babel, language became disconnected from reality, and the Adamic ideal was lost.

The importance of the Adamic view is today a matter of some controversy; some, like Hans Aarsleff and James Knowlson, see it as having been a central concern, whereas others, like Mary Slaughter, see it as having been less important.[20] In either case, though, there is agreement that by the early seventeenth century the question of the nature of language was becoming much more pressing. And in Bacon's *Novum Organum,* widely seen as one of the first works of modern science, language comes under as much fire as it does in works by those who appeal directly to the Bible as a source of authority.[21]

Bacon argues that "the subtlety of nature is greater many times over than the subtlety of the sense and understanding" (*Organon,* X). Unfortunately, those who study nature "falsely admire and extol the powers of the human mind [but] ... neglect to seek for its true helps" (*Organon,* IX). As a result, "the discoveries which have hitherto been made in the sciences are such as lie close to vulgar notions, scarcely beneath the surface" (*Organon,* XVIII).

The reason for the failure to develop a fuller understanding of nature is this: "one flies from the senses and particulars to the most general axioms" (*Organon,* XIX), so that "the axioms now in use, having been suggested by a scanty and manipular experience" (*Organon,* XXV) are really "*Anticipations of Nature* as a thing rash or premature" (*Organon,* XXVI).

The primary reason for the failure of people to attend carefully to nature, and to adopt what Bacon called an "inductive method,"[22] is that they are entranced by a group of "Idols," those of the marketplace, tribe, cave, and theater. Because of the effect of these idols, "the human understanding is like a false mirror" (*Organon,* XLI).

Most important are the idols of the marketplace:

> There are also Idols formed by the intercourse and association of men with each other, which I call Idols of the Market-place, on account of the

> commerce and consort of men there. For it is by discourse that men associate; and words are imposed according to the apprehension of the vulgar. And therefore the ill and unfit choice of words wonderfully obstructs the understanding. Nor do the definitions or explanations wherewith in some things learned men are wont to guard and defend themselves, by any means set the matter right. (*Organon,* XLIII)

Indeed,

> [T]he Idols of the Market-place are the most troublesome of all: Idols which have crept into the understanding through the alliance of words and names. For men believe that their reason governs words; but it is also true that words react on the understanding; and this it is that has rendered philosophy and the sciences sophistical and inactive. (*Organon,* LIX)

According to Bacon, who here launches into an attack on the then current Aristotelian science,

> The idols imposed by words on the understanding are of two kinds. They are either names of things which do not exist . . . or they are names of things which exist, but yet confused and ill-defined, and hastily and irregularly derived from realities. (*Organon,* LX)

There is, it should finally be noted, hope: the idols "must be renounced and put away with a fixed and solemn determination" (*Organon,* LXVIII). Only then can one escape the power of the "shadows thrown by words," can one avoid being, like Aristotle, "the cheap dupe of words."[23]

Bacon's attack on Aristotle provides an indication of the nature of his concern about language. It is the naming function of words, rather than, say, their more strictly rhetorical functions, that is of concern. But if we consider the examples that Bacon gives, we find that he is not being so generally critical as it might first have seemed. As "names of things which do not exist," he derides "Fortune, the Prime Mover, Planetary Orbits, Elements of Fire, and like fictions" (*Organon,* LX). But when we turn to those terms that are "confused and ill-defined," we find that there are "degrees of distortion and error. One of the least faulty kinds is that of names and substances" (*Organon,* LX). In fact, the problem with language appears fundamentally for Bacon to derive from the failure of people who use language to be careful and use the appropriate procedures. And, indeed, if we look to Bacon's understanding of the function of science, we see why language is not a problem:

> What I purpose is to unite you with *things in themselves* in a chaste, holy, and legal wedlock; and from this association you will secure an increase beyond all the hopes and prayers of ordinary marriages, to wit, a blessed

> race of Heroes or Supermen who will overcome the immeasurable helplessness and poverty of the human race.[24]

If the failure of science derives for Bacon from an improper method, the use of the proper method can achieve a traditional goal, one that can be "traced to Renaissance magical and alchemical sources," of "man's scientific domination of nature, and ... [equally] man as nature's 'servant and interpreter.'"[25]

To this goal, achieved through a "chaste and holy wedlock," language may appear irrelevant. Yet for Bacon language is strikingly important—in a way that separates him from Descartes and those who follow, and in a way that from a geographical point of view places humans on the other side of a great divide. From a contemporary point of view, his insistence on experiment and induction may appear to be a call for a science that is "theoretical" in form. But he was writing at the end of an alchemical and magical tradition, itself grounded in Aristotelian essentialism, within which the central aim was to divide the natural world into the correct set of categories; it was the failure to do so that, as we have seen, was the source of error. This categorization, for Bacon, involved the creation of what he called "tables."

A brief recitation of one of these tables, this one on "Instances Agreeing in the Nature of Heat," shows the problem:

> the rays of the sun, especially in summer and at noon; the rays of the sun reflected and condensed, as between mountains or on walls, and most of all in burning glasses and mirrors; eruptions of flame from the cavities of mountains; liquids boiling or heated; sparks struck from flint and steel by strong percussion; animals, especially and at all times, internally; strong vinegar. . . .[26]

The problem, of course, is that from what we might take as a systematic point of view this is a wildly improbable list. How would one get a handle on it? How would one remember it? Bacon had an answer that placed him squarely within the rhetorical tradition and particularly within its appeal to the art of memory.

Memory, as we saw earlier, was crucial in oral cultures to the maintenance and transmission of culture. But meter and rhyme, central in oral cultures, are of little help in the case of attempts to recall a list like the one just quoted. Fortunately, a solution to this difficulty was recognized at least as far back as 100 B.C., when it was described by Cicero in the famous story of Simonides. Simonides, the story goes, was hired by Scopas, to provide a eulogy for him at a banquet. He "recited a poem which he had composed in his praise, in which, for the sake of embellishment, after the manner of the poets, there were many particulars introduced concerning Castor and Pollux." Irritated, Scopas told Simonides that because one half of the poem concerned Castor and Pollux, he would pay only half of their agreed upon price; Simonides would need to get the rest from the other objects of his praise. "A short time after, they say that a message was brought in to Simonides, to desire him to go out, as two youths were waiting at the gate who earnestly wished him to come forth to them; when he arose, went forth, and found nobody." While he was outside, the building collapsed, and all inside were crushed; indeed, the bodies were mangled beyond recognition. But Simonides is said,

> from his recollection of the place in which each had sat, to have given satisfactory directions for their interment. Admonished by this occurrence, he is reported to have discovered that it is chiefly order that gives distinctness to memory; and that by those, therefore, who would improve this part of the understanding, certain places must be fixed upon, and that of the things which they desire to keep in memory, symbols must be conceived in the mind, and ranged, as it were, in those places.[27]

Here, according to tradition, Simonides invented—or at least formally recognized—the art of memory, a means for keeping in memory large numbers of apparently unrelated facts. And the art of memory does this through an appeal to *place,* not through rhyme and meter. The person who desires to remember a list of objects first imagines a large building, one with many rooms. Such a place, in fact, is the first thing that a person interested in developing this art must "construct." Then, the objects are "placed" within the building, in such a way that they would be encountered in the appropriate order during a walk through the building. To remember them then requires only that one mentally go on that walk; at the appropriate locations the objects, words, or names will become present to memory.

The art of memory, as Paolo Rossi has pointed out and as has been shown in greater detail by Frances Yates, was an essential feature of the practice of rhetoric through the Renaissance.[28] But as in the rhetorical tradition, it is an untheorized art; it exists alongside various theories of memory and the mind, but itself appeals to none. Rather, it retains its importance simply because it works, because it appeals to what its proponents have seen as a universal feature of places, that they embed themselves with great force on the memory. Now, Bacon has argued for a thoroughgoing critique of the language of science and for its replacement with a better one that does not have the inaccuracy, vagueness, and ambiguity that he found in the scientific language of his time. Yet he has remained within this tradition of the art of memory.

Descartes also wished to see a better language, a purely artificial language and perfect language. And yet despite some literature to the contrary, the perfect language of Descartes and those who followed is radically different from the improved language proposed by Bacon. It is different because the art of memory is excised from it. With it is excised the appeal to places, and with that the possibility of taking texts seriously. For in the art of memory the removal of places also removes time, and we are left with a written work that is read and remembered not by a metaphorical walk through a series of places but in an instantaneous act of apprehension.

In a letter to Mersenne, Descartes takes up the issue of the possibility of a perfect language.[29] He rejects a then current proposal, by Des Vallée, which Mersenne has introduced in an earlier letter. That proposal, he says, involves a scheme for both a new grammar and a new vocabulary. Agreeing that it might be possible to develop a language that is grammatically regular, and for that reason easier to learn than existing languages, he argues that the author of the plan in question has really provided no solution to the larger problem of how to develop a vocabulary that can be easily learned and is at the same time practical.

The problem, he suggests, is that the author of the proposal has not thought the matter through; the solution to the problem with vocabulary will not come, as has been suggested, from the appeal to some more generally acceptable set of words, perhaps a set derived from the historical study of language, and thus closer to some Adamic ideal. Rather, the development of a perfect language requires that one see that

> order is what is needed: all the thoughts which can come into the human mind must be arranged in an order like the natural order of the numbers. In a single day one can learn to name every one of the infinite series of numbers, and thus to write infinitely many different words in an unknown language. The same could be done for all the other words necessary to express all the other things which fall under the purview of the human mind. . . .
>
> [I]n any case the discovery of such a language depends on the true philosophy. For without that philosophy it is impossible to number and order all the thoughts of men, or even to separate them out into clear and simple thoughts, which in my opinion is the great secret for acquiring true scientific knowledge. If someone were to explain correctly what are the simple ideas in the human imagination out of which all thoughts are

compounded, and if his explanation were generally received, I would dare to hope for a universal language very easy to learn, to speak, and to write. The greatest advantage of such a language would be the assistance it would give to men's judgment, representing matters so clearly that it would be almost impossible to go wrong. . . .

I think that it is possible to invent such a language and to discover the science on which it depends: it would make peasants better judges of the truth about the world than philosophers are now. But do not hope ever to see such a language in use. For that, the order of nature would have to change so that the world turned into a terrestrial paradise; and that is too much to suggest outside of fairyland.[30]

The extent to which Descartes is rejecting the classical art of memory is made clear in his earlier *Cogitationes Privatae* (1619–21). There, speaking of a work on memory by Lambert Schenkel, he argues that

I thought of an easy way of making myself master of all I discovered through the imagination. This would be done through the reduction of things to their causes. . . . When one understands the causes all vanished images can easily be found again in the brain through the impression of the cause. . . . This is the true art of memory, and it is plainly contrary to his [Schenkel's] nebulous notions. Not that his art is without effect, but it occupies the whole space with too many things and not in the right order. The right order is that the images should be formed in dependence on one another.[31]

Now, Judovitz has argued that

Descartes's particular foundational interpretation of representation involves the culmination and also exhaustion of the Neoplatonic philosophical and rhetorical traditions of the arts of memory. . . . [I]ntuition is presented as a substitution and displacement of memory, and yet its schematic character as the reduction of metonymic chains to a causal and common figure relies on the same principles as the arts of memory.[32]

But, in fact, we find in Descartes the rejection of a fundamental feature of the art of memory, the use of places as ordering devices, and the introduction of a very different view, where causality and mathematics are the fundamental methods of ordering intuitions. Indeed, unlike the taxonomic theories of language proposed by Des Vallée and promoted in the seventeenth century by Wilkins, Comenius, and others, the view promoted by Descartes is a "mathematical, atomistic" language.[33]

He lays out the lineaments of this language in his *Rules for the Direction of the Mind* (1628). There he notes that "the exclusive concern of mathematics is with questions of order or measure" (*Rules*, Rule Four, 378). This conclusion suggested that it should be possible to discover a kind of universal mathematics, which deals with these issues in the most general way. Keeping this idea in mind, one needs to put things in the appropriate order, because

> all things can be arranged serially in various groups, not insofar as they can be referred to some ontological genus (such as categories into which philosophers divide things), but in so far as some things can be known on the basis of others. (*Rules*, Rule Six, 381)

Ultimately, "in order to make our knowledge complete, every single thing relating to our undertaking must be surveyed in a continuous and wholly uninterrupted sweep of thought, and be included in a sufficient and well-ordered enumeration" (*Rules*, Rule Seven, 387).

In one sense Descartes's discussion of language is unimportant; indeed, as we shall soon see, what is much more important in his thought is a vision wherein language is quite irrelevant. And yet in an important sense Descartes has laid out here *the* view of the nature of language in science. In this image language is a system of signs for internal thoughts. The signs are connected not through mnemonic devices, nor through the aims of rhetoric; indeed, "those with the strongest reasoning and the most skill at ordering their thoughts are always the most persuasive, even if they speak only low Breton and have never learned rhetoric" (*Discourse*, VI, 7). Rather, the signs are connected by the subject because they have mathematical or causal relations one with another.

The signs constitute a system, which in a fundamental sense can be mapped onto the ideas, and which has several fundamental features. First, it is a whole. The addition of new elements alters the entire structure of the system. Second, it is outside of time. It is unaffected by tradition. Most important, and this Descartes *does* take from his contemporaries who were devising universal languages, to learn the language is to acquire knowledge. In an appropriate language, it is not just that ideas have the correct names, or that general terms are appropriately defined. More important, the structure of the language maps onto the structure of the world. It is this view that has been most central to subsequent understandings of the language of science.

At first blush this seems an odd, even eccentric, view of language. But it has persisted well into the twentieth century, where it has informed that wing of the philosophy of language and of science informed by the work of Bertrand Russell. And I suggest that it is central to the reading of portions of contemporary geographic documents. As perhaps the most obvious example, take "the language of mathematics," as it is often described—and was in David Harvey's standard positivist text, *Explanation in Geography*.[34] It is not enough, if one is to become mathematically literate, merely to be able to compute mechanically the outcome of an equation. Rather, one needs to be able to *see* how the equation works, to see the relationships that define that equation. One needs, whether dealing with linear equations, logarithms, differential equations, or matrices, to be able to see the relationships among the elements, to see the equation as a structure. In learning the language of mathematics, one learns the nature of those relationships, ones of necessity, of covariation, and of indeterminacy. If one believes that mathematics is the language of nature, then to learn mathematics is at the same time to learn of the kinds of relationships that can exist in the world.

Similarly, and less abstractly, to learn the language of chemistry—of atoms and electrons and protons and valences—is at once to gain knowledge of the world, because within the language of chemistry well-formed statements as a matter of necessity map onto the world.

In the case of geography, we find appeals to several such artificial languages, and there too to learn such a language is considered to be to learn something about the world. Mathematics and statistics are perhaps the most obvious to some geographers, but the map is even more broadly used—and treated as having a language, the language of maps. Within the language of maps there are certain conventions of representation, of projection and scale and symbolization. Within the traditional theory of cartographic representation, it is possible on the one hand to develop an understanding of the nature of the ideal system of representation; this is what psychophysical studies of the perception of circle size and the like are about. On the other hand, once one learns the language of maps, to look at a map is at once to gain knowledge of the relationships that exist among the elements of the map, and is thence to learn about the world.

In the case of maps, of course, it is important to note that a recurrent criticism, that map symbolism is merely typological, a contemporary variant of Aristotelian essentialism, is the very same criticism that was lodged against many of the alternative language projects, of Wilkins and the like; these languages merely codify existing knowledge and cannot lead to new knowledge. Indeed, one of the fundamental features of Descartes's artificial language and of mathematics is that with them one can create new sentences that are well formed but far beyond what might have been imagined merely by "looking at the facts"; it is this very claim for mathematics that has so deeply informed the work of several people in geography, as in the case of Peter Gould and the advocacy of Q-analysis.[35]

Ultimately, although this view of language appears to make it a central issue in science, it has quite the opposite effect. Why is this? It is because this view establishes an ideal of a universal language that would in a sense be created out of the amalgamation of all existing (at least, accurate) texts. The statements within each individual text become merely instances within the whole language. Individual written works, as a consequence, come to be read only *for* those elements, and the rest of the text comes to be seen as inessential detritus. The result is that the written work ceases to be important, and it drops out of consideration.

MENTAL VISION AND THE NONLINGUISTIC TEXT

This view where language can be perfect or universal causes the individual text to lose importance as a result of the centrality of language, but in Descartes we find another view that appears to be almost the opposite of the first. This is the view that language is fundamentally unimportant to issues of knowledge, and hence of science. With this view the individual text comes to be unimportant for a very different set of reasons, but as different as the two views are, they have coexisted in contemporary understandings of the nature of science and geography.

Perhaps the simplest way to characterize the difference between the two views is this: In the first, language intercedes between the knower and the world, and language is a mode of expression of the ideas held by the knower. In the second, the mode of expression is considered to be unimportant, something that need not be perfected but rather can be excised from the process of knowing. In the first case there was a universal text; now in the second we have a nonlinguistic text.

To say this may seem odd, even preposterous; yet the notion that one needs to treat written works as nonlinguistic objects has been central to the understanding of science. Although the Platonic development of a kind of language suited for the written communication of ideas and the Cartesian development of a notion of the universal language have each been important at various junctures, the nonlinguistic text has had an overarching importance. It is in Descartes that we first find it formulated, and it is this formulation that has remained at the forefront.

Descartes's fundamental aim was to develop a means for making his way between the dogmatism of the predominantly Aristotelian science, current at the time, and the alternative skepticism expressed by Montaigne.[36] Montaigne had tried—as Descartes was to later—to develop a deeper self-understanding by isolating himself from the world and reflecting upon himself and his relation to it. The result, though, was not a systematic understanding:

> For Montaigne, the differences internal to the self are but the mirror of endlessly proliferating circumstances and events. His self-description cannot extricate itself from a description of the world.
>
> Self-knowledge and knowledge of the world mirror each other through the paradigm of interpretation, which cannot be fixed for lack of a foundational difference separating the subject and the world.[37]

By contrast, Descartes attempts a tack that will cut through the difference between knowledge of the world and knowledge of the self; in so doing he hopes to start anew.

The way in which Descartes attempted to move beyond skepticism to a more adequate knowledge is by now well known, but it will be useful to rehearse it to understand its relevance to the issue at hand. In his *Discourse,* he begins with the assertion that reason "is to be found complete in each of us" (*Discourse,* VI, 2). Still, "so long as I merely considered the ways of other men, I found little ground for assurance... [and so] I learnt not to believe too firmly anything that I had been convinced of only by example and custom" (*Discourse,* VI, 10). Having rejected the authority of custom, Descartes looked for a model of human knowledge. He concluded that "there is not such great perfection in works composed of several parts, and proceeding from the hands of various artists, as in those on which one man has worked alone" (*Discourse,* VI, 11); the best knowledge, then, was likely to be the result of a single person's efforts. Knowledge, he asserts, need not arise from a long, cumulative process; indeed, it would be possible to "reform my own thoughts and rebuild them on a ground that is altogether my own" (*Discourse,* VI, 15).

Looking back to the subjects that he had learned as a student, to philosophy, logic, and even mathematics, he concluded that although there was something to be learned from each, all were deficient in important ways. Hence, he determined to set up his own set of rules different from and better than the rules that characterized the reasoning in philosophy, logic, and mathematics. These rules were first, "never to accept anything as true if I had not evident knowledge of its being so"; second, "to divide each problem I examined into as many parts as was feasible"; third, "to direct my thoughts in an orderly way; beginning with the simplest objects, those most apt to be known, and ascending little by little"; and finally, "to make throughout such complete enumerations and such general surveys that I might be sure of leaving nothing out" (*Discourse,* VI, 18–19).

Having established this set of rules, Descartes then made the fundamental turn, as he determined, at least rhetorically, to doubt all that he knew, and then to start over. It is here, of course, that he rejects the evidence of the senses and asserts that all that he can be sure of knowing is that "I am thinking, therefore I exist."

Now, on the face of it this strikes one as a bit fishy, because it looks as though Descartes is deducing his existence through a process of reasoning. One is left with the image of a person running through a syllogism: anything that thinks also exists, I am a thing that thinks, therefore I exist. The problem is not so much that his argument seems fallacious. Rather, it is that it seems to assume so much already.

Yet to look at the argument in that way is to lose track of what is truly important here. And that is that the *cogito* argument is not really an attempt to deduce his existence. Rather, as Jaakko Hintikka has argued, it might better be seen, using a term introduced by John Austin, as a performative.[38] For Descartes, *in* the act of thinking one is existing. Moreover, in thinking, one is engaged in the human activity that can involve the achievement of certainty.

But for Descartes the appeal to the syllogistic reasoning formulated by Aristotle and used by his Scholastic teachers did *not* lead to this certainty. Rather, one needs for certainty to turn to intuition,

> [t]he conception of a clear and attentive, mind, which is so easy and distinct that there can be no room for doubt about what we are understanding. . . . [I]ntuition is the indubitable conception of a clear and attentive mind which proceeds solely from the light of reason. Because it is simpler, it is more certain than deduction. (*Rules,* Rule Three, 368)

Today we may think of intuition as a matter of guesswork, but Descartes had a very different image in mind: "We can best learn how mental intuition is to be employed by comparing it with ordinary vision" (*Rules,* Rule Nine, 400).

Hence, the central image of the process of thinking, insofar as it leads to certainty, is one of vision; one grasps the truth with the mind's eye. And this grasping involves the application of a kind of mental vision to the contents of the mind; the mind is represented here as a kind of interior space filled with ideas. Thus, "the whole method consists entirely in the ordering and arranging of the objects on which we must concentrate our mind's eye if we are to discover some truth" (*Rules,* Rule Five, 379).

This mental vision can be directed at all of the contents of the mind, which include memories, products of the imagination, sensations, and conceptions, although "it is of course only the intellect that is capable of perceiving the truth" (*Rules,* Rule Twelve, 411). But each of these is seen as the object of the mind's eye, and each is thus transformed in nature or function.

So, for example, the products of imagination are no longer seen as "images," pure and simple. That is, they are no longer seen as necessarily resembling that of which they are copies:

> [O]ur mind can be stimulated by many things other than images—by signs and words, for example, which in no way signify the things they signify.... [W]e must at least observe that in no case does an image have to resemble the object it resembles in all respects.... Indeed the perfection of an image often depends on its not resembling an image as much as it might. (*Optics,* VI, 112–13)

In an important sense the products of the imagination no longer need have spatial characteristics—and this shows the extent to which Descartes's notion of thinking, as a mental vision within a mental space, differs from one derived strictly from common sense. Similarly, sensations are not, as in Aristotle, seen as associated with the body; rather, they imprint themselves on the mind itself. Hence, speaking of a piece of sealing wax, Descartes in the *Meditations* asserts that "the perception I have of it is a case not of vision or touch or imagination—nor has it ever been, despite previous appearances—but of purely mental scrutiny" (*Meditations,* VII, 31). But the new vision of the mind is nowhere more important to later understandings of the nature of the written work than in the matter of memory, for it is the problem of memory that leads to a theory of deduction and that is fundamentally related to the excision of time from understanding. With the excision of time the notion of mental vision is made complete, and the text becomes merely, at best, a helpful device.

The question of memory comes up in the most important way in the issue of deduction, which in turn arises as we try to get beyond the very simple. Descartes asserts that a major source of error is the failure to look beyond complex situations, and to decompose them into their simplest elements. The key to knowledge, he argues, is that we can have a clear and distinct apprehension of the simple elements of which the world is composed; as long as we keep our attention turned to them we shall not be fooled. As he put it, "whatever method of proof I use, I am always brought back to the fact that it is only what I clearly and distinctly perceive that completely convinces me" (*Meditations*, VII, 69).

The problem, though, is one of combining together those individual perceptions. Although it may very well be that I can clearly and distinctly perceive this and now that, how do I go from individual perceptions to draw warranted conclusions and to construct theories? For Descartes this move was full of danger because it is so easy for a group of perceptions, all once clear and distinct, to be combined into something that is far less so. This problem meant that "in order to make our knowledge complete, every single thing relating to our undertaking must be surveyed in a continuous and wholly uninterrupted sweep of thought, and be included in a sufficient and well-ordered enumeration" (*Rules*, Rule Seven, 387). And yet he immediately recognized a problem: "For this deduction sometimes requires such a long chain of inferences that when we arrive at such a truth it is not easy to recall the whole route which led to it" (*Rules*, Rule Seven, 387). Hence, the very possibility of a real, that is, a certain, science rests on resolving the problem of memory, for "whenever even the smallest link is overlooked, the chain is immediately broken, and the certainty of the conclusion collapses" (*Rules*, Rule Seven, 388).

The solution, he argues, is this:

> Since memory is weak and unstable, it must be refreshed and strengthened through this continuous and repeated movement. . . . That is why it is necessary that I run over them again and again in my mind until I can pass from the first to the last so quickly that memory is left with practically no role to play, and I seem to be intuiting the whole thing at once. (*Rules*, Rule Eleven, 409)

The solution, then, is to understand that knowledge need not proceed through a series of steps. Rather, it is possible to have a complete grasp of a situation, to have it all in mental view at once. Knowledge, when viewed as a kind of mental seeing, does away with the problem of memory, the problem of sequence, and the problem of time. It does away, that is, with reading. One who truly knows the nature of something knows it all at once.

In fact, this understanding of the nature of knowledge extended beyond the mere laying out of currently accepted knowledge, and extended to the matter of proofs; as Ian Hacking has written,

> for Descartes a proof was a device that enabled a man to pluck scales from his eyes and see the truth. An angel, for example, with perfect "mental gaze"

> would require no proof. . . . Descartes . . . thought proof a device for getting rid of words, enabling a man to perceive the connections between ideas steadfastly.[39]

This Cartesian understanding of the nature of knowledge has another consequence: it makes knowledge utterly private. Words do not have a direct relationship to ideas; they are connected with them only through what Hacking calls a relationship of precedence or consequence. That is, a word may conjure an idea, or an idea conjure a word, but the relationship between the two is not necessary; at best it is causal and contingent. And so, I have an idea and a word comes to mind, I utter the word, another person hears it, and an idea is conjured in that person's mind. Here the public discourse of words is a poor imitation of the certain, internal discourse of ideas. Words can at best provide signs that I have knowledge but can never truly demonstrate that knowledge.

Finally, if knowledge is private, it is also best understood as an individual matter. In knowledge, as elsewhere,

> there is not usually so much perfection in works composed of several parts and produced by various different craftsmen as in the works of one man. . . . [T]he sciences contained in books . . . do not get so close to the truth as do the simple reasonings which a man of good sense, using his natural powers, can carry out in dealing with whatever objects he may come across. (*Discourse*, VI, 11–13)

Conclusion

Descartes has laid out a powerful and influential view of the nature of knowledge. Knowledge is an individual, personal matter. It resides within the human mind, which is separate from the physical world. It consists of ideas, rather than words. It is a matter of grasping ideas and their relationships in a kind of interior, mental space. And this grasping can occur all at once; we can have simultaneous knowledge of the world.

This view pervades contemporary understanding of the sciences; indeed, each element must appear strikingly familiar. It is common to judge a person rather than that person's work, and to compare individual works with "what the person knows." It is common to think that we can appropriately say that a person "knows" this or that, that a person understands a structure even if he or she has never described it. It is common to see linguistic missteps as just that, and as irrelevant to the person's understanding of the nature of things. It is common to imagine knowledge as having a determinate structure. And it is common to believe that one can grasp that structure all at once. All of this leads directly away from an interest in the written work. Reading, concerned with words, sequential, appealing to authority rather than experience, seems to provide a model for all that is improper, for all that leads from knowledge to opinion. What is to be idealized is the individual, ready at once to face worlds, the internal and the external.

This view is in contrast to that of "the other Descartes," and that was a view, in support of a universal language, that would seem to value texts highly. For the other Descartes our everyday language is faulty but might be replaced by another better able to capture the nature of the world. This language would be mapped directly onto the world, so that to learn the language would be at once to gain knowledge of the world. Like the view of the other Descartes, this view has been widely held but in a more specific form. It is held by those who use mathematics, the periodic table—and maps.

But there too are individual texts devalued, because universal language schemes promote the notion of a universal *text*. The universal text would be a compendium of all written works, which would consolidate all knowledge. But this image suggests that any individual text needs to be seen primarily as a repository of elements to be mined for the larger project. Individual texts themselves are of no interest.

Finally, in the development of views on the nature of language and texts in science, there is a third, equally important view, that advocated by Plato. The importance of Plato lies in the way in which he expressed an alternative to earlier oral means for the transmission of knowledge. His alternative used the written work as a repository, as a means for the transmission of knowledge in the form of abstract conceptions not easily amenable to oral transmission. Moreover, Plato's work expresses a valuation of the process of criticism, based on the view that it is the practice of criticism that leads to knowledge.

Yet here, too, there is an underlying tension. If in one sense we can now ask "What did Plato say" about this or that, and if there is a scholarly industry devoted to answering just those questions, in another sense both the content and the form of his written work tended to undercut that project. The content, a theory of knowledge that saw knowledge as constantly battling to keep from being ensconced as merely true belief, was contrary to the idea that one could properly say what Plato had said. And the form, that of the dialogue, provided a model very different from that of the written text. In the case of Plato things are even more complicated. Although his written works and their reception have tended to value the text, and the industry of interpretation that has grown up around his works has further supported the notion of individual texts as valuable objects of scrutiny, these projects have ironically led attention away from a central point—that the very give and take of his works, the dialogical nature, remains in scientific texts today.

These three themes have been central to the contemporary understanding of the written work. Indeed, Foucault can speak of the death of classicism and the rise of philology as expressive of a greater attention to language, and Hacking can speak of the time when language went public, but for the vast majority of those using language in science the news has not arrived. Written works store knowledge in a retrievable form, we can imagine and sometimes use a universal language that gives us an entrée to the real, and scientific knowledge is private and nonlinguistic. The ambiguity and ambivalence inherent in the expressions of these views in Plato and Descartes are effaced, and each is in its turn appealed to. The written work, as an individual object, as a singular attempt to represent something about the world, disappears.

CHAPTER TWO

Formalizing Common Sense (II): The Work in the System

We saw in the last chapter that the written work in geography is typically viewed as the medium through which the geographer represents the world. It is through the written work that we are exposed to a particular version of the world. But it is also through the written work that we learn something about the author/geographer.

As we saw, the traditional image of the geographical work is suspicious of language, seeing its everyday use as fraught with difficulties. And these difficulties seem especially pronounced if we take as ideal an image where a purified language maps directly onto the world, without the mediation of social groups. Here we have a system where the geographer's mind contains an image of the world, which maps to the written work, and then the mind of the audience, from whence the audience may see the world as did the originator of the ideas. And here the written work has tended to slip away, to be effaced, in favor of a "universal text."

This purified image of geography and science may seem far removed from the everyday practice of science, but in fact, the two are closely linked. The codification of the traditional common sense has proceeded on parallel courses; just as the understanding of the issue of representation has gradually elided the boundary between place and space, coming to see the ideas of science as existing in a grand conceptual space, so too has the understanding of the various more worldly functions of the written work turned away from the particular and the place. Indeed, it seems clear that in the case of the understanding of representation, important elements of the written work have been occluded by a powerful set of images, and that this has also happened in other areas.

Here, too, a common sense, a set of homilies that entreat the aspiring scientist to do this and not that, has been simplified into an image of a social system, an image that is compelling in its purity. Here, though, the process of purification seems to have been occasioned by a series of social changes. Indeed, the image that has guided thinking about representation has its roots in the seventeenth century, but the sources of the image of science as a social order are very much newer. We can see the development of this image in a range of places in this century. We see it in the creation of a new sociological theory about science, where there had been none before. We see it in the development of systems for the classification of knowledge. And we see it in the gradual conceptualization of what it means to own, to have rights to, and to be responsible for a geographical work. At the same time, we see it in the work itself, in the rise of manuals for scientific publication, in the development of standardized systems of citation and citation analysis, and in the emerging fears of fraud and the consequent creation of codes of professional ethics.

We saw in the case of representation that the power of the images of a science with a universal language or with no language at all obscured a central feature of science, the reliance on dialogue and critique, and we saw that that reliance was there almost at the beginning, because the roots of Western science extend to the Greek polis. Similarly, in the case of the social order of science there is a related obscuring; the appeal to the image of structure obscures the central role of community in science.

In this way, as in others, both images are decidedly modern in the appeals that they make to images of space; in the case of both representation and the social order of science, a notion of space is implicitly adopted, a notion that both restricts and obscures the nature of everyday practice. Indeed, it restricts the ability of the student to see the geographical nature of that practice. These issues will become clearer in the next chapter, where I turn to critiques of the standard images; in what follows I shall restrict myself to an elaboration of this second, and seminal, image of geography.

The World of Written Works

Perhaps the most obvious feature of the written work in geography is that it fits into a well-established typology, an established set of categories. Most obvious, of course, are the map and the journal article. The history of the map has been well recited; the journal is perhaps slightly less well known. The journal, a periodical compendium itself made up of smaller units, has existed for about three hundred years. In 1665 the initial issues of the first journals, the *Journal des Sçavans* in France and the *Philosophical Transactions* in London, were published. The first journals replaced earlier means of communication among scientists, like the erudite letter, which had itself flourished after the establishment of the state and the postal service. The erudite letter shared some of the features of modern journals; designed to be read aloud at meetings, it typically was written in an impersonal style, and in fact, there were individuals who became central to the dissemination of these letters, and hence in a sense filled the function of modern publishers. (The erudite letter was succeeded for a time by an intermediate form, the manuscript newsletter. The primary difference between the two was that the newsletter was more a commercial venture.)[1]

The erudite letter accommodated elements of older face-to-face communication to a new institutional and technological era, and the new scientific journals went even further. Seeing their main functions as providing news and establishing priority in scientific discovery, they were both formal and commercial. According to Douglas McKie, the *Journal des Sçavans,* for example, was expected to contain "details of new books, obituaries, news of experiments in physics and chemistry, [and] discoveries in the arts and sciences. . . ."[2] Somewhat more scholarly in the present sense, the *Transactions* was designed to present material of interest to professionals, rather than to a wider group of amateurs; it presented original research results, and in that way can be seen as a better model for the modern journal than was the *Journal des Sçavans.*

This century, and especially the years since World War II, has seen a real efflorescence of journals. Derek J. de Solla Price in fact used the establishment and growth of the journal as a fundamental indicator of the growth of science. He noted that by 1961 about six million articles had been published, and every year one half million were added to the pile; further, the number of journals itself was doubling about every fifteen years, so at the then-current rate it could be predicted that the fifty thousand existing in

1961 would by the year 2000 have increased to one half million.[3] And in fact, in 1995 the library of the University of California at Berkeley reported holdings of 349,728 serials. This massive growth in the number of journals has established the journal article and abstract as the fundamental constituent of science.

And so in three hundred years science has moved from a system of communication that was largely local, and face-to-face, and within which ideas were seen as a matter of course as being held by people in places, to a system within which there are so many works, so many ideas, that one feels compelled to treat the output of science as being very much like the output of any large-scale industrial process. There, too, one feels justified in seeing those publications and ideas as engaging in a process of motion not simply through an ideational space but also through a set of other spaces, of commodities and classification systems.

The System of Authority

The notion that science is in part a social system that can be talked about systematically is remarkably new. Its early genesis is usually traced to Ludwik Fleck in the 1930s,[4] but in the English-speaking world it is probably more accurate to say that it developed later, with Robert K. Merton's famous monograph on Puritanism and science.

Merton developed what has turned out to be the central approach to the sociology of science in the twentieth century, and at the same time he in a very real sense defined the "standard" system, the standard image of science.[5] Fundamental to his approach are two sets of notions. First, science is seen as constituting a seamless web, where it has been possible to imagine the practice and the content of science as intimately related. And second, scientists are believed to hold a basic set of norms or values, which, indeed, members of society more broadly also believe to be true of science. Those values are communism, universalism, disinterestedness, and organized skepticism.

With respect to the more general image of science, Merton argued that scientific practice, the linguistic and symbolic expressions of scientific work, and the social hierarchy of science are somehow structurally homologous. On his view the mapping between the practice of scientific research and the linguistic and symbolic expressions of that work within the system of communication is indifferent with respect to direction; one can as well deduce the world from a written work as the written work from the world.

Similarly, by using citations, references, and the concept of intellectual ownership as a fulcrum, it has been possible to move from the written work in the other direction, to the level of hierarchy and rewards, and to deduce the quality of a person's publications from the rewards that he or she has received, just as to deduce the rewards from the quality of the work.

For Merton the three "levels" of science are interrelated in a way that renders science a functioning whole, one that has provided a uniquely successful means of knowing. This relationship works because scientists hold a set of values that in a sense tie it together. The first of these is communism, which concerns the attitudes of scientists toward their work. The facts and findings of individual scientists are expected to be made publicly available, and to be generally knowable and manipulable. Because scientific works have an unusual characteristic—when one "uses" the works of others one does not thereby "use them up"—it is possible for a scientist to get credit for findings, while at the same time, "the substantive findings of science are . . . assigned to the community."[6]

The second ideal, universalism, refers to the desire of scientists to appeal to universal standards. In the case of social scientists this may lead only to modest, "meso-level" explanations, but the goal is nonetheless to move away from the particular. Because it appeals to a term like "universal," this goal puts science in the camp with other universalizing activities, like myth and religion. Here this ideal has an overt evaluative element, because it suggests that the truths attained by scientists are in some way "better" than those that are the result of more particularistic enterprises, like history, just as it seems to suggest that scientists reside "higher" on a hierarchy where the quality of knowledge is the overriding concern.

The third ideal, disinterestedness, implies not an attitude held by individuals but an institutional structure within which those individuals act. To say that such an ideal exists is to say that science is itself disinterested, in the sense that it is composed of a set of structures that form an apparently neutral surveillance system.[7]

Finally, organized skepticism requires the establishment of a set of rules according to which one is deemed appropriately skeptical, and it involves its own set of assumptions about when and why a body of assertions and assumptions is to be deemed consistent and supportable.

According to Merton, this basic set of values can be said to guide scientists in their work, although from his point of view it does not matter, actually, whether scientists really hold those values or merely act as though they do. What matters is that most scientists act in accordance with them;

as long as they look as though they, by and large, are seeking the truth or are being dispassionate in their peer reviewing, we can continue to act as though science is driven by those values. Scientists adopt these four norms, perhaps knowingly, perhaps not, as they become socialized into science.

And the important thing to note about these norms is the way in which they are related both to an individual's place within the scientific community and to the work that that person produces. On the one hand, the adoption of these values fits neatly into a cast of mind that works—or sees itself as working—through an open-minded, skeptical process, one fundamentally receptive to new and counter-evidence. In that way these ideals support what is generally described as the "scientific method" of gathering of evidence, hypothesis formation, testing, and so on.

On the other hand, these values point to a way in which hierarchies are established within science. Although communism is seen as an ideal, at the same time built into science is a strong belief in intellectual "property rights," and in the necessity that scientific "discoveries" be made public. Hence, the accrual of intellectual property is the means by which scientists gain authority. Correlatively, because this property is communal, to the extent that individuals create property the body of science is itself enhanced. Perhaps most important here, because the fundamental means of the accrual of rights to intellectual property is by making that property public, via publication, the processes of discovery, publication, and the creation of hierarchies within science can be seen as structurally homologous and interrelated.

Now, this view of science has been subject to a wide range of criticisms. In fact, one might see the current sociology of science as a series of footnotes to Merton. Some have argued that Merton's list of norms is incomplete. Some have said that we need to see these norms as existing in a tension with a set of "counter-norms."[8] And some have said that the norms, are in fact simply ideological devices, used to prop up the position of science more generally.[9] Yet there remains something compelling about Merton's picture; it appears to capture something about the commonsense perception of the nature of scientific practice, and especially about scientific practice in the age of the journal, and as we shall see, the age of the citation.

Classification

If we turn to a second way of thinking about the written work, as an element of a system of classification, we find much that seems familiar. It would be pointless to attempt to argue that systems of classification of the written work are new, and indeed, this is not what I wish to argue. There are systems for the organization of knowledge in Plato's *Republic*, in Aristotle's *Metaphysics*, and in Zeno. By A.D. 305 we find the famous system of Porphyry (see figure 1).

This system was repeated in later works, and as Ernest Cushing Richardson points out, strikes us today as "a most interesting suggestion of the modern evolutionary order."[10] Moreover, as the modern age began, there was a flowering of such systems of classification; they were developed by Bacon, Descartes, Hobbes, Vico, Kant, Schelling, and Hegel—to name only a few.[11] But although such systems were extremely common, it remains that they were not widely used in the cataloging of books.

Indeed, Melvil Dewey created his famed Dewey Decimal System in 1876 just because there was such an anarchy of systems; it appeared that almost every library had its own.[12] And in an important sense, what distinguishes the newer systems from the older ones is this: neither the French system, the Dewey system, nor the Library of Congress system (1901) can really be seen as a system based—as were those of Aristotle, Porphyry, Bacon, and Kant—on the organization of knowledge. Rather, they were practical systems, which ordered books themselves in ways meant in the end only to render their use less difficult. We find much the same when we look to systems developed for the cataloging of maps, although there the logic of organizing maps by region—and by size—imposed itself early enough that it is rare to find attempts to develop conceptually elegant, or even coherent, systems.[13]

In part because of these classification systems, today the circulation of books occurs within a highly organized system. But this is at the same time a physical system; books are circulated within libraries, and among libraries. There are local, regional, and national libraries, just as there are ones associated with government, universities, and businesses. The purchasing and lending practices of these libraries are highly differentiated, and to choose to write a certain sort of book, to write for a certain audience, to write in a certain language, to choose a particular publisher, all act to place one in a particular way within these systems.

With respect to cataloging, whether an author likes it or not, his or her work is, simply because of the way that it was written and published, made a part of several cataloging systems.[14] These ordain that in every library that uses a given system, the work be placed in a certain place. Within this

Substance
- Spiritual
- Corporeal
  - Celestial
  - Terrestrial
    - Elementary
    - Mixed
      - Lifeless
      - Living
        - Vegetable
        - Animal
          - Irrational
          - Rational
            - Man

Figure 1. Porphyry's system of classification

system, of course, it is suggested that all books cataloged under B are more alike than they are like those cataloged under G, and so on. Further, those cataloged BD, BG, and so on are presumed all to share some general features, and yet to be fundamentally different. So if this latticelike structure, fundamental to both of the standard American systems, the Library of Congress and the Dewey Decimal, is not truly based on a scheme for the organization of knowledge, it nonetheless assumes that there is a homology between classificatory systems within libraries and the practical needs of those using books. And so, in writing a book and (in the United States) registering it for copyright protection, one implicitly accedes to this hierarchical view, and to the ontology that accompanies it. One accedes to this view in which there is no privileged place; rather each volume is simply an element in a pragmatic hierarchy.

Ownership

We find much the same modernist structure expressed in the way in which books and journals and maps are commodities, and commodities like any other. Of course, in another sense they are *un*like any other, and in fact, the commodity form of the text is rather new. The written work came to exist first as a physical object, and only later did it begin to be seen as a set of ideas behind the object. Indeed, this rethinking involved a reconceptualization of both author and work, where ultimately what developed was a view wherein both author and work can best be seen merely as aspatial armatures, to which are attached sets of rights and responsibilities.

When we think of the ownership of geographical works today, the matter seems straightforward. The objects are indeed seen as dual in form. On the one hand, they consist of the actual physical expressions of ideas. These expressions are legally protected through the law of copyright. The protection, though, is temporally limited, and at the same time is alienable; indeed, in articles and more in the case of maps, the author or creator most often sells his or her rights to a publisher, who may sell them again. On the other hand, the geographer also owns the ideas themselves. If I develop a theory or method, it is mine. This form of ownership, of course, is very different from the first, because it is permanent and cannot be alienated. My work is always mine, for better or worse. Now, this dual form of existence of the scientific and geographical object is a commonplace; it is something about which we seldom feel the need to comment. But there are two things to notice about it. First, and as I suggested earlier, it is historically contingent, and rather new. And second, recent developments have begun to recast it.

The relationship between the written work and the author has been considered in a number of recent works, perhaps most notably by Michel Foucault, Mark Rose, and Martha Woodmansee.[15] Rose's and Woodmansee's works, the first about Britain and the second primarily concerned with developments in Germany, are both concerned with the ways in which the notion of the author developed, and with the relationship between that notion and the notion that the author is the producer of a physical and an intellectual work. Both direct their attention to the eighteenth and especially the early nineteenth centuries as the eras in which these notions were formalized and codified into law. And both directly address the ways in which formal, legal structures interacted with technical and social changes to produce a series of conceptions of the physical work, the intellectual work, and their interrelations.

At the same time, both of these works address the period well after the invention of printing, and well after an important event to which Foucault, albeit briefly, turns his attention. Granting that in the early part of the period discussed by Rose and Woodmansee the modern category of "literature" had yet to be invented, and that the writing of scientists was therefore not seen as disciplinarily distinct from that of others, Foucault points out that it would be a mistake to stop there. In fact, in the era before the invention of printing, and until the invention of modernist science, the scientist was in a sense the author par excellence; that is, the very fact that something had been written by Aristotle, to take the obvious example, guaranteed its truth. So in the premodern era the scientist-author had a special sort of authority, which only came to be criticized successfully with the rise of modernism.

At the same time, the rise of modernism was the event that gave rise to a new form of the author—and of the work. But this did not happen easily. Certainly key in the development of this form was the printing press. Invented late in the fifteenth century, it immediately made possible a fundamentally different method for the dissemination of knowledge. Whether in the case of the book or the map, it now became possible to produce large numbers of identical objects. Publication produced written and graphic objects that could be easily bought and sold. In addition, it now became possible to engage in textual criticism, and thereby to compare sources, with the view to creating the "best" text.[16] Because books and maps were manufactured objects, their "titles changed from addresses to the reader to become like the labels on boxes."[17] Their very proliferation necessitated the invention of systems by which to catalog them. For the same reason, because they were among the first items in the "age of mechanical reproduction," they began to lose what Walter Benjamin referred to as their "aura."[18]

Perhaps most important, though, in the case of both books and maps, it was clear that the production of such objects required the joint collaboration of a wide range of people. Books required editors, typesetters, printers, bookbinders, and booksellers; maps required engravers and printers. And as today, the labor required of each was not insubstantial; Arthur Robinson estimates that in the case of maps produced by copperplate, an expert engraver might produce one square inch per day.[19]

This division of labor was reflected in the earliest regulations of the ownership of books. They consisted of grants, by governments, to individual booksellers; those grants gave a bookseller a monopoly within a geographic area over a certain category of works. The earliest copyright law, England's 1710 Statute of Anne, was intended to do the same; it was promoted by booksellers who saw their investments in typesetting at risk from literary pirates. What was protected was the right of the bookseller to labor invested in a physical object. Indeed, although now seen as the first copyright statute, there is a sense in which the Statute of Anne was more concerned with physical than with intellectual property.[20]

Still, and notwithstanding the intent of its framers, the statute came to be interpreted in a way that gradually has led to a more recognizably modern view. As Rose has noted, in England this involved two distinct steps. First, it was necessary that the rights to a work come to be seen as residing—at least in the first instance—in the author rather than in the bookseller. And second, it was necessary that the physical object come to be seen merely as a set of signs for the "real" work, which had only a virtual existence.

The first step happened quickly. Indeed, one of the very first suits under the Statute of Anne, concerning the translation of a work by a scientist familiar to historians of geography, geologist Thomas Burnet, pressed that point. And by 1735, when a bill for improving the statute was introduced, it had become clear that "a significant evolution had occurred, in which the focus of the literary-property question shifted from the bookseller to the author."[21]

In England the second step occurred at about the same time. Rose argues that its first manifestation is in a lawsuit brought by Alexander Pope in 1741. In that suit (*Pope v. Curll*), Pope argued that he maintained the literary right to letters that he had written, and therefore physically relinquished. In the court's decision, as Rose put it, "the author's words have in effect flown free from the page on which they were written. Not ink and paper but pure signs, separated from any material support, have become the protected property."[22]

The two steps were not, of course, unrelated. In part this is because the statute and the common law in which it was embedded were expressive of what after Locke's "Second Treatise" came to be termed the "labor theory of property." Locke argued that when people mix their labor with portions of the natural world, that portion of the world becomes theirs.[23] Here, as is well known, Locke has assumed that people can be seen as owners or proprietors of their own selves.[24] But the crucial point is the way in which this view tied in with another view, and one equally popular in the empiricist Britain of the eighteenth century. On that view the mind presents itself to the world as an empty slate. The external world quite literally impresses itself on the mind, and thereby creates impressions. These, in the mind, can be manipulated, as ideas.

In striking contrast to what since Freud has come to seem common sense, for empiricists like Locke and Hume it is possible for one to read one's own thoughts, very much as one would read a book, in the visual way seen in Descartes.[25] Here, then, the creation of a written work is at the same time an act of labor that writes that book into (or onto) one's own mind. So one owns one's own ideas in a way more fundamental than the way in which one owns some external object, or to take a favorite case, a piece of land.

This connection between the ideas that constitute the work and the person is closer still in the romantic tradition. In the eighteenth century the understanding of writing in terms of adherence to a traditional body of rhetorical rules, an understanding that until then had coexisted uneasily with a notion of writing as motivated by the muse, dropped away.[26] What remained was the muse in a more modern incarnation, in the form of the notion of genius. Here it became not labor but internal mental qualities that were the source of the work. As Woodmansee has described the matter,

> moments of inspiration move, in the course of time, to the center of reflection on the nature of writing. And as they are increasingly credited to the writer's own genius, they transform the writer into a unique individual uniquely responsible for a unique product. That is, from a (mere) vehicle of preordained truths—truths as ordained either by universal human agreement or by some higher agency—the *writer* becomes an *author* (Lat. *auctor*, originator, founder, creator).[27]

Indeed, by the end of the eighteenth century, Johann Gottlieb Fichte, putting a decidedly neo-Kantian cast to the matter, remarked that

> each individual has his own thought processes, his own way of forming concepts and connecting them. . . . Now, since pure ideas without images cannot be thought, much less are they capable of representation to others. Hence, each writer must give his thoughts a certain form, and he can give them no other form than his own, because he has no other. But neither can he be willing to hand over this form in making his thought public, for no one can *appropriate* his thoughts without thereby *altering their form*. This latter thus remains forever his exclusive property.[28]

So in an important sense, by the end of the eighteenth century the physical work of the writer, or the scientist, had lost its aura. But to a certain extent that aura had migrated to the work as an ideal object of ownership and to the author as an owner. Within science this was expressed in some areas in the recognition that there are certain ideas that are clearly the property of those people who discovered or created them; in other areas we see it in disciplines like history and archaeology, where there is a strong sense of one's work as being based within a particular set of facts. In both cases it is common to believe that both the scientist and the ideas deserve special consideration.

At the same time, this romantic view has within the context of an increasingly modernizing and globalizing economy been threatened by an alternative, a Lockean view that attempts to dispense once and for all with the view that we can treat either the author or the idea as special.[29] Codified most forcefully within Anglo-American theory of copyright, this view presses for a system of rights and responsibilities to written works that can in the first instance be operationalized in specifiable ways. For example, the ideal is that rights be of limited and specified duration and that ownership be alienable, in a way that can be clearly delineated within a written context.

It is important here to see that the Lockean view of property developed on a parallel course with a view of the individual. In describing the acquisition of the right to an object as the result of one's having mixed one's labor with that object, Locke appealed to a metaphor of substance. Yet the upshot of that conception was to see the relationship between an individual and that object as one in which with the ownership of the object the individual acquires certain rights and obligations. Indeed, ownership ultimately comes—and especially when what is owned is a mass-produced object—to be seen more and more as only a matter of the holding of such rights and obligations.[30]

Nowhere has this been more clear than in a recent controversy about journal prices. Reacting to high prices for journals, several individuals and associations have published analyses of journal pricing and "cost-effectiveness." For example, Henry Barschall, a physicist, compared journals in terms of their cost per one thousand characters, and also in terms of each journal's cost per citation generated by its articles.[31] One publisher, Gordon & Breach, appeared on both accounts to be extremely expensive and not cost-effective. The publisher's reaction was simple; it sued the author. In a related matter Gordon & Breach procured an injunction prohibiting the American Mathematical Society from mentioning its journals in cost surveys. It (apparently) under a fictitious name attempted to gather evidence from librarians about the effects of these articles on their decisions to purchase periodicals.[32] And finally, it sued a University of Southampton chemist because a year after leaving the editorial board of its *Molecular Crystals and Liquid Crystals,* he started a competitor, titled *Liquid Crystals.* In one sense all of this should come as no surprise; Gordon & Breach argued that "[academic] societies, which are generally non-profit and have tax-exempt status, can consistently price at a lower rate; membership dues, advertising, government grants and other forms of support insure that their subscription prices will be lower than those of a commercial publisher,"[33] but they failed at the same time to note that "the Association of Research Librarians earlier this year reported that publishers' profits had increased from 40 to 137 per cent from 1973 to 1987."[34]

The effects have been clear, as when

> University [of California] lawyers told Joseph A. Boisse, the university librarian at U. C. Santa Barbara, "it would be prudent" to retract an article in a faculty newsletter that decried price gouging by the publishers because they had threatened to sue. Boisse cited Gordon & Breach's *Early Child Development and Care,* up from $462 per year in 1987 to $1,080 in 1988.[35]

Here, too, we see the development of a conflict, between the Mertonian structure, in which one can read status from the publishing record, and where publishing, in turn, can be mapped onto the quality of the author's ideas, and the current practice of publishing articles (and books and maps) within a commodity system, where the nature of authorship is reconceptualized in a more Lockean way, and where the pressures of the market are in various ways "distorting."

Conclusion

And so if we look at three very prominent ways in which the written work has been characterized, and in which the common sense about the written work has been formalized, we find a recurrent pattern. Whether the work is looked at as an element in the social system of science, as an element of a larger system of classification, or as a piece of intellectual property, the formalization of the commonsense understanding has been in terms of a structure, and a structure that is fundamentally modernist in form. This, indeed, is very much what we should expect on the basis of what we saw in chapter 1, because there we saw the extent to which such structures had come to constitute the commonsense way of looking at knowledge and at the world itself, where the world has become a system, a picture.[36]

At the same time, it would be a mistake to imagine that a conceptual scheme characterized in the works of philosophers and scientists has by itself had the power to force into line the entire world of administrators, catalogers, booksellers, and publishers. And in fact, here we need to see the two issues, the representation of representation and the representation of the work in the world, as being much more closely tied one to another. We need, that is, to see the ways in which the existence of the written work has itself acted in support of that view of knowledge, just as a view of knowledge has supported the production of the work.

In what follows we shall see just how strong those ties are, as we shall see the difficulties that have attended attempts to recast the nature of knowledge, difficulties that in large measure arise as one attempts to rethink knowledge in ways that the written work does not allow. And we shall see that fundamental to these difficulties are the ways in which the written work comes to be involved in the construction and maintenance of a wide range of places. At one scale, we call one of these the "modern world," but those involved in publishing are at the same time creating particular kinds of places at other scales, as are those who read, who write, who cite, and who enter into critique.

PART II

Beyond the Traditional Common Sense: The Author, the Work, and the World Beyond

CHAPTER THREE

Formulating the New Common Sense

In the past several years the conventional wisdom about knowledge, in geography and elsewhere, has come under a series of sustained attacks. Although these attacks seem to have focused broadly, to the extent that at times it has appeared that it is being claimed that the discipline ought simply to scrap everything and start over, at the same time they have consistently involved a dual claim about the written work. It has been asserted that substantive written works have not been what their authors claimed or believed. And it has been asserted that writings *about* writing have been equally inadequate, to the extent that they have failed to recognize the problems with substantive works.

This may sound like a tidy situation, but it is not, for several reasons. For one, some geographers have begun to argue that landscapes themselves can be seen as texts, so in a sense there *are* no purely substantive works, and *everyone* is writing about "the written work." And others have argued that language is just fundamentally inadequate to the needs of representation, and that *no* written work is more than an expression of feeling, a local response to a local situation. In the first case we are asked to look at everything as a text; in the second we are told that once we do so, all is lost.

To be honest, within geography very little has been written about the written work—at least if we leave aside those periodically appearing normative statements by journal editors and association presidents. And this lack is in stark contrast to the situation in other disciplines, like anthropology, where there has been an outpouring of works about the written work. It is also in contrast to the situation more broadly in the history of science and the sociology of science, both of which have become much more fundamentally and critically interested in the written work than they once were. Finally, we have begun to see a substantial body of work on science emerging from literary studies and from a revivified rhetoric. Together these are beginning to constitute, or at least to attempt, a new common sense.

Because these analyses of the written work have been articulated from within a variety of disciplinary traditions, they emerge in patterns and structures that vary greatly. But—and at the risk of being criticized for privileging a single of those traditions—I would argue that we can see this new common sense as having three main moments. First, it has an ontological moment; claims are made about what does and can exist. Second, there is an epistemological moment, as claims are made about the nature of knowledge and of its possible foundations. And third, claims are made about the nature of language, and particularly about the extent to which it is or can ever be robust enough to represent what we know about the world.

In characterizing this extraordinary variety of works in this way I am, as I suggested, taking some liberties; from the point of view of someone in literary studies or rhetoric the list may simply seem at best the wrong way about, and at worst entirely too pat. But in describing it in this way I am not (necessarily) suggesting that everyone else uses (or ought to use) the same categories; rather, I am offering this simply as a list of areas where substantial questions recur.

I suggested that there has been a relative silence on the written work from within geography, in contrast to the din without. But it is nonetheless the case that this new common sense about the written work has had an expression within geography. There, indeed, we find among advocates and critics of postmodernism the elaboration of a series of themes that are very much of a piece with the themes considered by those in the various sociological, historical, and literary studies of science. And so it will be useful in laying out this new common sense about the written work to attend at the same time to the arguments made within geography on the postmodern challenge.

In what follows I shall first lay out the lineaments of the new common sense, which has been elaborated theoretically within geography, but individual elements of which are widely found, both in the discipline and outside. I shall then turn to three areas in geography—the new cultural geography, the history of cartography, and feminist geography—where the written work has been the subject of scrutiny. Finally, I shall turn to some recent criticisms of the new common sense and shall there set the stage for what follows in chapter 4, a more broad-based attempt to understand and transcend the limitations of this new common sense.

Formulating the New Common Sense

WORDS CUT LOOSE FROM THE WORLD

A central area of concern among many postmodern geographers is language, and it is common to see the assertion that we are in a crisis of representation, where the meaning of terms seems fluid, disconnected. In part this crisis has involved the recognition that in geography, as elsewhere in science, not all language is literal, that language is also used figuratively. And in fact, the issue of the role of metaphor in geography has been studied in a variety of ways by recent geographers, who have looked at the relationship of metaphor to place, to theory, and to the history of the discipline.[1]

It has been traditional to see metaphors, at the very least, as involved in the formation of concepts in science; we saw a recognition of this as long as forty years ago, in works by philosophers like Max Black and Norwood Russell Hanson.[2] There, though, it was possible to imagine that one used metaphors in the process of discovery and then dropped them in favor of more literal, and scientific, language. And certainly for those who followed Black in arguing that a metaphor was more than a mere comparison, but ineluctably involved an element of tension and uncertainty, it seemed that the move from the metaphoric to the literal was essential if the product was, truly, to be science.[3]

More recently some have argued for a broader role for metaphor, seeing it is a kind of heuristic device, and one useful for enabling the student of a new subject to transcend older ways of thinking.[4] Going perhaps a bit further, and echoing work of Steven Pepper, the recent work of George Lakoff and Mark Johnson has argued that language far more generally is structured in terms of "big metaphors," like battle and war; they and their followers have argued that we need to attend to the ways in which those metaphors shape our understanding of all that we do.[5] And, indeed, references to Lakoff and Johnson's work have appeared in a variety of contexts within the geographical literature.[6]

But according to some, matters are more serious; it is not simply that we need to see that geographers now and again use metaphor, for discovery or as a heuristic device or even as a broad strategic tool. Rather, as Gunnar Olsson said, we must now recognize—as he believes many have—that while to

> tell the truth is to claim that a = b and be believed when you do it . . . after Nietzsche everybody knows that a in fact is *not* b. . . . And thus it is that meaningful truth contains an element of lying, for every truth is formulated around a nucleus of difference.[7]

Here Olsson refers to Nietzsche's "On Truth and Lie in an Extra-Moral Sense," and with good reason; this text has been seen by many as a fundamental expression, even a fountainhead, of a central tenet both of postmodernism and of the studies of the written work. And here Nietzsche is taken to be saying that all language needs to be seen as fundamentally figural, that notwithstanding claims that this or that assertion is attempting to get at what is factually and literally there, the reality is that language is never literal. Every assertion contains elements of the figural; metaphor is everywhere. And indeed, this does appear to be a view that we find in Nietzsche:

> What, then, is truth? A mobile army of metaphors, metonyms, and anthropomorphisms—in short, a sum of human relations, which have been enhanced, transposed, and embellished poetically and rhetorically, and which after long use seem firm, canonical, and obligatory to a people: truths are illusions about which one has forgotten that this is what they are; metaphors which are worn out and without sensuous power; coins which have lost their pictures and now matter as metal, no longer as coins.[8]

Here, according to a common interpretation, the whole idea of a language that is crisp, clean, and well defined goes by the boards. And so, too, must much that we call science. In the Cartesian past, according to Olsson, it was believed that there was a "book of nature," and that our scientific knowledge was doubly "anchored" in it and in the "invisible" ideas that constitute our theories.[9] Our knowledge was seen to be an attempt to grasp or capture that reality, and language was the medium through which that capturing occurred. The crisis of the sign has occurred because it is no longer believed that signs, or language more generally, have the ability to reflect adequately that reality. Rather, they are now seen as both arbitrary and volatile.

Closely related to the notion that language is fundamentally figural has been the argument for the importance of rhetoric to science. Now, this might mean a number of things. For one, there have been a series of works, like Charles Bazerman's *Shaping Written Knowledge: The Genre and Activity of the Experimental Article in Science,* Alan G. Gross's *The Rhetoric of Science,* and Lawrence J. Prelli's, *A Rhetoric of Science: Inventing Scientific Discourse,* that have been rhetorics of science in the strict sense.[10] Even in these works, however, one finds wide disagreement. In Bazerman, as in earlier works such as Joseph Gusfield's pioneering "The Literary Rhetoric of Science: Comedy and Pathos in Drinking Driver Research," the aim is to understand the persuasive elements of the works of science, but at the same time to assume that the understanding of those elements will enable us better to understand what is left over, the truths of science.[11] By contrast, Prelli's interest is in the creation of a "topological logic" of the rhetoric of science. That is, his primary aim is to look at the process of rhetorical invention, that process of the creation of persuasive arguments.

Like the earlier students of the use of metaphor and other figural language in geography and science, Bazerman and Prelli can be seen as basically leaving science alone; Prelli, for example, allows that "I do not intend to argue that everything in science is rhetorical, without remainder."[12] Gross, by contrast, takes a more sweeping view:

> [W]e can engage in a systematic examination of the most socially privileged communications in our society: the texts that are the chief vehicles through which scientific knowledge is created and disseminated. We can argue that scientific knowledge is not special, but social; the result not of revelation, but of persuasion.[13]

As he put the matter in a review of Bazerman,

> [I] want to enlist rhetorical analysis in the service of radical epistemological conclusions.
>
> ...the only way to draw a principled line between the rhetorical and the non-rhetorical in science is to press rhetorical analysis to its limits, to see whether there are any limits.[14]

And indeed, this is the point at which we can begin to see the connection between postmodernism's claim that all is figural and its related claim that all is power. For taken to its conclusion, this is what Gross and others are prepared to conclude, or have already concluded. Where Bazerman and traditional rhetoricians have been willing to imagine it possible in a practical, technical, and effective way to analyze the operation of texts—in the same way that one might analyze the operation of any machine—and to prescribe means to make them work more effectively, postmodernists and other students of science, including constructivists and discourse analysts, have argued that one can never find a spot with a purchase firm enough to allow one to begin. For them the rhetorical, like the figural, "goes all the way down." As John Bender and David E. Wellbery said,

> *Modernism is an age not of rhetoric, but of rhetoricality,* the age, that is, of a generalized rhetoric that penetrates to the deepest levels of human experience. The classical rhetorical tradition rarefied speech and fixed it within a gridwork of limitations; . . . Rhetoricality, by contrast, is bound to no specific set of institutions. It manifests the groundless, infinitely ramifying character of discourse in the modern world. [emphasis in original][15]

In the modern world, *all* language turns out to be persuasive; the scientists' dreams of a distinction between the demonstrations of science and the persuasions of the rhetor were simply that, dreams.

Whereas the claims that language is always figural and always fundamentally concerned with the process of suasion have seemed to undercut the traditional common sense about science, there has been a third element of the new common sense that has gone further. The claim for the intertextuality of scientific texts has seemed to make the meaning of any text fundamentally unknowable.

Put as simply as possible, intertextuality is the view that "any text is constructed of a mosaic of quotations; any text is the absorption and transformation of another."[16] There is, of course, a sense in which this notion has long been more a commonsense one in the case of science than of the novel. Indeed, it has at least in this century been common to recognize formally this intertextuality through the process of footnoting, and the notion of intellectual authorities long precedes the formal footnote.

Moreover, it is equally a commonplace that individuals read texts in different ways. For some, for example, David Harvey's *Explanation in Geography* should be seen as his crowning achievement. For others it is simply the work of a young scholar who has not found his way, and his later *Limits to Capital* shows a more mature force. For yet others his *The Condition of Postmodernity* is the work to be reckoned with. Each of these readers would place the favored work in a different relationship with the others; each would see the favored work as residing in a different place on a career trajectory. At the same time, even readers and admirers of a single text vary substantially; for someone grounded in philosophy much of *Explanation* will seem straightforward, even textbookish, whereas the discussions of mathematics and modeling may seem quite new and challenging; for someone whose background is in modeling the opposite may seem the case. Further, the way in which a text is read depends on the intentions and place of the reader, whether a potential reviewer, a critic from an opposing camp, out to find fault at every turn, or a struggling undergraduate. And each of these relationships will almost certainly be altered over the course of time; what seemed sharp and clear in retrospect seems banal, while insights not noticed come to the fore.

All of this, as I have suggested, is quite commonplace; we tell our students these things and expect them, gradually, to understand what we mean. But there is more at issue here, especially in light of the claim that language is fundamentally figural. For *that* claim seemed to remove the possibility of fixing accurate referents to terms; it suggested that when we refer to something, the class of objects to which we are making reference is never as clearly defined as we might like. And this additional claim suggests that the sources of linguistic authority may be much more far-flung than we had thought, and that as a consequence there can be even less certainty about what we mean when we make a claim, even in what on the face of it appears the most spare and deliberate text. Now, it seems, the number of possible readings of any work has increased beyond limit. An author cannot be sure what her audience will bring to the text, and they cannot be sure what *she* brought.

So there really are three ways in which there has recently developed a new common sense about the language of science; any work is now argued by many to be figural, rhetorical, and intertextual. Now, it might be countered that the average person's understanding of language has little to do with the obscurities of postmodernism or of Julia Kristeva or of rhetoric. But I would also suggest that a bit of reflection will show that this is not really the case. The recognition that metaphor is all around us is one that lately seems to need little argument; we hear about metaphor in all branches of geography. The notion that texts are written to persuade is implicit, or explicit, in most style manuals and works written for aspiring geographers. And the notion that texts are strongly interconnected is at the heart of recent discussions about computer networking, hypertext, and the like. All that has been missing is a way of systematically ordering these insights.

ON WHAT THERE IS

Closely tied to these views of language is an altered view of ontology, of views about what exists. According to Olsson, the move from modernism to postmodernism has involved a move from questions of epistemology to ones of ontology:

> [W]hereas modernism deals primarily with epistemological questions of a relational nature, postmodernism experiments with ontological issues of fundamental importance. As modernism searches for purposeful answers to well defined riddles, postmodernism explodes in cascades of paradoxical situations.[17]

Here again we see what looks like a real sea change. During the quantitative revolution, geographers, acting in accordance with what they saw as an ideal of science, had attempted to diminish the size of the group of "objects" that were counted as real. They took almost as gospel the need to see the world in terms of a simple set of objects and processes, so that a theory that took as basic only the individual and distance seemed the ideal. Alternative approaches such as the regional or the humanist, which were content with a world of great diversity, were viewed with a kind of revulsion, seen as evidencing a lack of resolve.

Postmodernists have come full circle; for them the category of the "real" has, as Olsson put it, "exploded." This means that a whole range of "things" can be taken to be real, and basic. We no longer need see ourselves as living in a world of objects and facts; events and processes can be real, and so too can places and localities and even epochs—the modern and the postmodern.

The theoretical underpinnings of this view have for many derived from the work of the American neopragmatist Richard Rorty, and his particular version of the history of philosophy has been taken as *the* history, as an objective representation of past events and ideas. Rorty has argued that we have reached the current stage via a series of steps. First, in the Enlightenment, and especially after Kant, we began to see the centrality of human experience to the construction of reality. Then, in romanticism we began to believe that there are different languages through which to have access to or characterize reality. Finally, since Nietzsche and the original pragmatists at the end of the nineteenth century, we have come to see language and knowledge as merely means of getting what we want. Here language goes "all the way down"; there simply is no prelinguistic reality.[18]

As a result, Michael Dear concludes, we see that "our conceptual orderings . . . do not exist in the nature of things, but instead reflect our philosophical systems."[19] Philip Cooke puts the matter similarly, when he declares that there is a "more fragmented, negotiated reality."[20] This conclusion seems, in a clear-cut way, a challenge to those who would believe that the world, and the objects or facts of which it is composed, exists in some simple way "out there," waiting to be found, and hence appears to be a rejection of much that has been assumed in geography and elsewhere to be true about knowledge of both the natural and the social worlds.

Postmodernism has itself been viewed by many with a combination of horror and disdain. For many of the more scientifically minded geographers, it has seemed no more than an antiscientific antirationalism; few of these have even bothered to comment in print about what they see as an aberrant phenomenon. At the same time, for others it has seemed too much a position developed in the comforts of an office. For Richard Symanski, for example, it bespeaks an unwillingness to get out of the office and into the field; it is at base unempirical.[21] And for some feminists it is the same old story of, as Doreen Massey strikingly put it, "flexible sexism."[22]

Together these critiques may seem to provide ample evidence that postmodernism is at the very least marginal. But it seems to me that here, too, to adopt this view is to be extreme. Indeed, here too it is the case that what looks marginal is very much, these days, a matter of common sense. Put for the moment most simply, the postmodern claim that the world consists of a wide variety of irreducible elements strikes me as capturing neatly a basic feature of contemporary experience, that the demise of traditional patterns of authority, whether in religion or science or politics, has left in its wake a feeling that at the very least many questions about what is "basic" are open to dispute. After all, in order to give credence to the postmodern claim about what exists, one need not give in completely; one need only allow that within certain arenas what was once taken to be settled now is not. And much that I have said about the written work points to this very fact. The development of citation indexes, for example, can be seen as evidence of uncertainty. Eugene Garfield's claim that citations provide a "real" means for determining the structure of knowledge, in contrast to the faulty ones provided by professional catalogers and by self-attributions, displays this same sense that the written work is itself an ambiguous object, claiming to be this when it is really that.

LOCAL KNOWLEDGE

Having said this, I should add that the discussions of what in geography and social theory is termed "ontology" all rely quite heavily on analyses of language, on the one hand, and knowledge, on the other. Indeed, and notwithstanding assertions to the contrary that "ontology is primary," most postmodern claims about what exists can best be seen as consequences of prior decisions about the nature of knowledge. And it is in the matter of epistemology that we begin to see the development of claims that seem to move far beyond what most would see as common sense. Whereas in the case of language and ontology the claims of postmodernists and of other students of the written work can be seen as appealing in some way to everyday experience, in the matter of epistemology we quickly find ourselves in the company of propositions about knowledge that few but scholars would be willing to entertain, let alone accept. This situation, it will turn out, exists because in epistemology, as nowhere else, the traditional version of common sense is both so deeply rooted and so strongly undergirded by institutional and technological supports.

Here one typically sees a series of related claims. First, and most closely related to the issue of ontology, is the claim that knowledge cannot, in the end, ever rest upon firm foundations. We recall that Descartes was in a line of philosophers and scientists who have sought in various ways to provide just such a ground. Bacon, for example, offered a scheme in which experiment provided this ground. To reject the possibility of firm foundations is not, of course, necessarily to reject the possibility of knowledge. In a sense, for example, the work of Karl Popper, rejecting the possibility of ever truly verifying results but claiming instead that at best we can only fail to falsify them, is itself a rejection of this common, Baconian form of foundationalism, which sees knowledge—both of facts and of concepts—as somehow capable in principle of directly mapping onto the world. Yet to the extent that the core of foundationalism is the justification of knowledge claims by reference to that which is directly evident, Popper, to the extent that he relies on notions of reason and induction, can be seen as equally involved in a foundationalist project.

In contrast, the rejection of foundationalism in an important way returns us to Plato's appeal to dialogue and place, to his claim that knowledge can only arise in the context of an ongoing process of argumentation. And indeed, the most popular version of antifoundationalism, propounded by Rorty, is very much in the Platonic, dialogical tradition. As he put the matter,

> pragmatists would like to drop the idea that human beings are responsible to a nonhuman power.... Pragmatists would like to replace the desire for objectivity—the desire to be in touch with a reality which is more than some community with which we identify ourselves—with the desire for solidarity within that community.... On this view, to say that truth will win in such an encounter is not to make a metaphysical claim about the

> connection between human reason and the nature of things. It is merely to say that the best way to find out what to believe is to listen to as many suggestions and arguments as you can.[23]

This view that all knowledge is ultimately grounded within social groups leads to a second claim, that all knowledge is, as Clifford Geertz termed it, irreducibly local.[24] Here too there are a number of possible ways of understanding this assertion. It might mean simply that individuals and groups hold beliefs that are specific to them. To say this seems no more than to suggest that we ought to look at the ways in which people order their lives, and on one reading this is to do no more than to validate the sort of work that geographers have been doing for many years, work on the ways in which groups of people perceive and order the regions and landscapes in which they live.

But the view most often associated with Rorty's claims is a more extreme version of this view, in which, quite literally, *all* knowledge is local. On that view, propounded most vocally by advocates of the "strong program" in the sociology of science, even science and even mathematics and logic must be seen merely as expressions of local practices and customs.[25] And this notion of local knowledge is often closely tied to one or both of two disturbing claims, that all knowledge is relative and that all systems of knowledge are incommensurable. Both seem to suggest that no claim to knowledge is any better than any other, and that in fact no one can understand anyone other than a member of his or her own community. As Dear put it, "The essence of the postmodern answer is that all such claims are ultimately undecidable."[26]

One may, of course, see Dear's claim as wildly improbable, and as one that even he does not really hold; indeed, his occupies an extreme on a continuum of types of relativism.[27] Yet even here, where common sense seems to have been left far behind, there is at least one area of familiarity: it has become a commonplace within certain of the humanities and the social sciences to see the characterizations of phenomena developed within different disciplines as "different but equally valid." Although few would argue that all geographical accounts of economic restructuring are equally good, there certainly are those who would see those accounts developed in economics, geography, and political science simply as different expressions of differing perspectives.

And so, we see over the last few years the development of what I shall call the "new common sense," a set of views about language, ontology, and epistemology that attempts to put the written work in its place (see Table 1).

Table 1. The new common sense

	Traditional common sense	New common sense
Language	Literal	Figural
	Demonstration	Persuasion
	Referential	Intertextual
Ontology	Coherent	Fragmented
	Single	Multiple
Epistemology	Foundational	Antifoundational
	Universal	Local
	Absolute	Relative
	Translatable	Incommensurable

This new common sense has gone a long way in cutting through what has surely been a failure to attend theoretically to what "everybody knew," that written works are ambiguous and complicated, that their authors attempt to present themselves in the best light to the right people, and that the field of science is crowded with competitors. But although this new common sense has much to recommend it, it has nonetheless remained tied too closely to the very views that it has attempted to overthrow. Exactly why this is will be more clear if we look to geographers' attempts to develop a "better" text.

The New Common Sense and the Written Work

Over the last several years there have been a number of attempts to put this new common sense into practice, to get beyond the traditional images of representation and structure, and to develop an understanding of the written work that will be theoretically more satisfying, while perhaps at the same time giving some credence to the insights expressed in common-sense accounts of everyday practices, insights that these traditional images have tended to obscure. In geography, though, we find very little with which to work. Granted, if we turn back to the 1960s, we find Symanski's attack on historical geography, and especially the work of Donald Meinig, as examples of "the manipulation of ordinary language."[28] But far from attempting to develop or work within a new common sense, Symanski was digging in his heels, arguing that it is indeed possible to speak plainly and literally. As Symanski pointed out, there had been one work, *Birds in Egg* by Gunnar Olsson, that had taken a more critical stance toward language.[29] Still, although widely admired, this and Olsson's later works were written more as general challenges to the practice of the discipline than as specific analyses of particular geographical works; for that reason, if not others, they have generated little sustained interest in the nature of the written work.[30]

Although work on the written work in geography was late arriving on the scene, by the late 1980s it had begun to appear in at least three areas. In human geography a series of authors writing within what is loosely termed the "new cultural geography" have focused on the use of figural language, and especially metaphor. In the history of cartography there have been the first concerted attempts, inaugurated by Brian Harley, to see the ways in which maps might implicitly have persuasive elements. And in the area of gender a series of authors, in the process of developing feminist geographies, have implicitly and sometimes explicitly based their claims that geography is currently gendered on the analysis of written works; in doing so they, too, have argued that the written work embodies persuasive elements and expresses relations of power.

Writing about the development of the new cultural geography, Denis Cosgrove and Mona Domosh were explicit in their claims that the new common sense offers a more adequate means for understanding the written work in geography.[31] Using as their foils works by Cole Harris and Susan Hanson,[32] they argued that "the types of writing advocated by Harris and Hanson do in fact serve to naturalize and therefore mystify the role of the geographic author in the former and the geographic text in the latter."[33]

Both Harris and Hanson, in fact, fail to understand that "the modernist tenets of value neutrality, the uniformity of nature and the experimental method are historical creations of specific time periods, cultures, and social formations."[34] Indeed, what we need is to recognize the ineluctability of this fact; we need to write texts in self-conscious ways that will make it clear both to ourselves and to our readers that we are writing from a point of view:

> In shifting the metaphor from landscape as system to landscape as theatre for example, we become conscious precisely of the metaphor as metaphor. . . . With the shift to more cultural metaphors we have been forced to abandon the innocence of representation, for we know that our metaphors are themselves drawn from the arena of human meaning creation.[35]

Ultimately, then, the problem is not that our ways of representing the world are fundamentally faulty, and that we need radically new ones.[36] Rather, we simply need to be explicit in our understanding of the ones that we, and others, are in fact using. The problem lies in the text, and in our relationship to it.

In a series of papers on the development of economic geography, Trevor Barnes has made similar arguments.[37] In keeping with the new common sense he argues that yes, figurative language is everywhere. Nonetheless, he does not go the distance with those who have argued that because *all* language is figural, there is no literal. Rather, drawing from an analysis of metaphor by Rorty, itself based on an analysis by Donald Davidson, he claims that metaphoric statements are distinct from literal statements because the latter have truth value and the former do not.[38] In contrast to the literal, the metaphoric is important for what it does,

> and that use is one of changing our beliefs through the jolt, the *frisson*, that a novel metaphor can produce....
>
> Metaphors, precisely because they are patently false and absurd, cause us to stop and think, and thereby possibly lead us to do different things than we have done in the past.[39]

Notwithstanding this positive feature of metaphors, there are, according to Barnes, problems with their use: most important, they can come to be seen as literal. And indeed, this is what has happened to the gravity model; initially taken strictly as a metaphor, and in a way that kept visible its limitations, it gradually came to be taken literally. So the important thing—and here he agrees with Cosgrove and Domosh—is to recognize the dangers that inhere in the use of metaphors, and to attempt always to keep at the forefront the fact that they *are* metaphors.

When Barnes speaks in this paper of the use of metaphors, this use is always within a text, and indeed, his concern is only with the text. In a more recent paper, on deconstruction and the quantitative revolution, he makes a similar argument but extends its scope.[40] Considering the rise of quantification in geography, he argues that it was adopted because it gave the promise of a "final vocabulary," which would after all provide a real groundwork or foundation for geographical knowledge. Arguing that mathematics was conceived by geographers as universally true, a model of logical inference, objective, simplifying, and precise, he attempts to show that there are good reasons for seeing every one of those attributions as inaccurate. He bases this argument on work in the sociology of mathematics by David Bloor, which itself derives from philosophical work by Wittgenstein.[41] And so, for example, he argues that

> the final claim about mathematics by the quantifiers is that it is precise, and unambiguous. The paradox here is that the only means of sorting out those standards of precision and unambiguity is by using ordinary language. For precision and unambiguity are not inherent in mathematics itself, but are

> criteria that are applied after the fact, and are thus debated and discussed in the vernacular.[42]

Barnes, though, does not follow up on this comment, and here, as in Cosgrove and Domosh, we are left with an analysis of the geographical work that focuses almost exclusively on language within the work itself. Perhaps more importantly, we are left at sea with respect to the most fundamental of questions. Can we truly understand one another? If not, how is it that we manage to make our way through our everyday life with so few mishaps? But at the same time, why do we seem so frequently to be misunderstood?

Beyond the New Common Sense

THE NEW HISTORY OF CARTOGRAPHY

A related argument was developed by Brian Harley, in a series of papers published just before his death. Harley's richly empirical work pointed toward the need to situate the work in the world, to see not just the author and the subject but also the wide range of economic and social and political contexts within which it exists.[43] Although to put it in this way is to exaggerate a bit, it is helpful to see "Maps, Knowledge, and Power" as a catalog of the places where one can find power relations in a map and "Silences and Secrecy" as a case study, "Deconstructing the Map" as a theoretical statement of the nature of the map, and "Cartography, Ethics, and Social Theory" as an attempt to place "Deconstructing the Map" in a broader context.

Harley began as a historical geographer, not a theoretician, and his work on the nature of maps in one sense suffers—but in another benefits—from his having gotten a late start at theorizing. Indeed, although he claimed that his work was grounded in a reading of Foucault and Derrida, Barbara Belyea has made a compelling case that he has misread them.[44] Nonetheless, it seems equally likely that with respect to the conclusions that one might draw in a particular instance, the three would probably have much with which to agree. And there the agreement would be in moving beyond the lineaments of what I have termed the new common sense.

Harley begins by noting that even critical students of representation in other areas have commonly subscribed to what I termed traditional common sense, where the map is seen as a mirror or direct representation of the world. On that view the errors in maps can be attributed to a failure of technology; in fact, we know *how* to map the world accurately, and given the time we shall do so. According to Harley, the power of this view has been seen as deriving from the visual nature of the map, the way in which it relies upon technology, and its pragmatic success, in getting people from here to there. Yet this view has obscured the fact that maps are commonly produced by institutions that have great power. Indeed, on reflection we find that "the scientific rules of mapping are, in any case, influenced by a quite different set of rules, those governing the cultural production of the map."[45] In fact,

> cartographic discourse operates a double silence toward this aspect of the possibilities for map knowledge. In the map itself, social structures are often disguised beneath an abstract, instrumental space, or incarcerated in the coordinates of mapping. And in the technical literature of cartography they are also ignored, notwithstanding the fact that they may be as important as surveying, compilation, or design in producing the statements that cartography makes about the world and its landscapes.[46]

Indeed, Harley continues, we would do better to replace the image of the "map as mirror" with that of the "map as text." In doing so we could begin to see that

> the issue in contention is not whether some maps are rhetorical, or whether other maps are partly rhetorical, but the extent to which rhetoric is a universal aspect of all cartographic texts.... Again, we ought to dismantle the arbitrary dualism between "propaganda" and "true," and between modes of "artistic" and "scientific" representation as they are found in maps. All maps strive to frame their message in the context of an audience.... All maps employ the common devices of rhetoric.... Rhetoric may be concealed but it is always present, for there is no description without performance.[47]

Ultimately, because maps are so good at concealing their own rhetorical nature, they are especially useful for those groups that wish to dominate.

In the two "empirical" essays, "Silences and Secrecy" and "Maps, Knowledge, and Power," Harley fleshed out this argument. In the first, beginning from the assertion that "that which is absent from maps is as much a proper field for enquiry as that which is present [because]... silence should be seen as an 'active human performance,' "[48] he laid out the use of secrecy and censorship in cartography in the sixteenth and seventeenth centuries in Europe. But he continued with an analysis of what he termed "unintentional silence." There,

> the silences in maps act to legitimize and neutralize arbitrary actions in the consciousness of their originators. In other words, the lack of qualitative differentiation in maps structured by the scientific *episteme* serves to

> dehumanise the landscape. Such maps convey knowledge where the subject is kept at bay. Space becomes more important than place: if places look alike [on the map] they can be treated alike.[49]

Harley's work has been widely praised, viewed as expressing the now obvious truth about maps. Many have seen it as ushering in an era of new awareness among historians of cartography, who had previously been doggedly empirical, and all too closely tied to those technical colleagues and allies who saw maps simply as mirrors of reality. And Harley's work has in fact been followed by a series of works in the same vein, like Denis Wood's *The Power of Maps*.[50] What these works have in common is an image of the map where it exists at the center of a set of interests, dominated by the interests of powerful groups. Those groups control the content of maps, by structuring what is both included and excluded. At the same time, because the map carries with it the aura of scientific respectability, it manages to convince even those in power that it is merely a neutral and nonrhetorical reflection of what is really there. According to this view the map is fundamentally rhetorical, like everything else, but has been better at appearing otherwise.

These recent works on cartography go beyond the new common sense in an important way, and beyond the work-as-idea view that we find in parts of the new cultural geography; they do this by turning attention to the situatedness of the map, to the political, social, and economic contexts within which maps are created. Nonetheless, they are only very vague in their explications of the ways in which in particular cases this situatedness leads to the creation of particular kinds of works. And rendering matters more confusing, one still finds in the works on maps, as in the new cultural geography, the explicit claim that *all* works are rhetorical, without an accompanying attempt to understand what this means for the person making such a claim. If all claims are ultimately groundless, why accept that one? To the extent that one sees the work as idea, those practices of working within an academic setting, the very practices that some have seen as providing a bulwark against relativist indecision, fall away.

ON FEMINIST GEOGRAPHIES

There is, though, one area within geography where there has been for a number of years both an explicit concern with the nature of representation and a vigorous debate about just these questions, of how works are generated and of how one might understand knowledge when the universalism of the traditional common sense has been undercut. In the area of feminist geography many take a stand that confronts at the outset the issue of relativism, while for obvious reasons including at the very outset an appreciation of the need to confront the issue of representation.

Although work explicitly on the written work is relatively new in feminist geography, there is one obvious sense in which feminist geographers (and nongeographers) have long appreciated the power of language, and that is in the matter of sexist language. Indeed, it is just such an appreciation that has been expressed in pressures for changes in the style of the written works, for the excision of sexist language, whether it refers to people or to objects. But although to many the need for linguistic change has been obvious, because—to take a single example—the use of masculine pronouns to refer both to men and to women seems so clearly exclusionary, an explicit theory of the role of language in structuring social relations has seldom been articulated in much detail. Rather, it has typically been left to the reader to make the connection. And so, alongside the explicit appreciation of the need to write in certain ways, there has been little in the way of the development of a theory about the role of the written work itself.[51]

Yet the extent to which the issue of language was implicitly part of feminist theorizing is made clear by the rapidity with which as soon as it was explicitly raised (in the late 1980s), a rich body of work developed. And this work, right from the outset and notwithstanding substantial differences among those concerned, involved the joint recognition of the situatedness of authors, of the place of the written work in a social system, and of the power of systems of representation in ways that made its analyses far more useful than those more firmly in the camp of the new common sense.

One portion of this literature has focused on the discipline of geography itself. For example, noting that women have largely been written out of even the most recent histories of the discipline, Domosh has pointed to the processes through which this occurred. Denied the opportunity for both academic training and traditional fieldwork, Victorian women nonetheless engaged in forms of research. Yet "women travelled, then, for quite specific reasons, but what they were seeking was as much empowerment and self-knowledge as 'objective' knowledge."[52] And so, the writings that resulted from this research were doubly resisted, as works by the unqualified on subjects of no interest.

A second portion of this literature focuses on much the same issue, the exclusion of women, but in the present. Doreen Massey's "Flexible Sexism,"

Rosalind Deutsche's "Boys Town," and Steve Pile and Gillian Rose's "All or Nothing? Politics and Critique in the Modernism-Postmodernism Debate" have taken mainstream economic geographers and postmodernists to task for reasons directly associated with, and deriving from, an understanding of the written work.[53] Pile and Rose, for example, write that David Harvey's *Condition of Postmodernity*

> is typical of many geographers' rejection of discourses which dare to challenge the security of their knowledges.... The theoretical centring of one causal process is sexist and racist, and the tyranny of white masculinity shows in the power invested by these texts in their unspecified, disembodied authors.... This abstract, distanced, cool voice of reason claims objective neutrality but actually reflects all those qualities deemed masculine by Western patriarchy including violence: the "epistemic violence" of exclusion.[54]

So here there is something about the very style of representation that is being used by Harvey to enhance his authority.

At the same time, Harvey has failed to recognize his own situatedness with respect to gender or race; his "account of postmodernism depends on feminism's absence."[55] And as a result, he has failed to see that "modernism and postmodernism are not the polar opposites which they are claimed to be, for the point of feminist critiques of both is the same: these discourses avoid the politics of gender and race."[56]

And finally, the book itself exists in a social system, as an element of its author's authority, but also of a broader set of structures. As Massey has pointed out, *The Condition of Postmodernity* is a kind of chit, used to certify that its author is part of an in-group:

> Thus Soja refers to Jameson "Fredric Jameson, perhaps the pre-eminent American Marxist literary critic" (page 62). Jameson repays the compliment: "that new spatiality implicit in the post-modern (which Ed Soja's *Postmodern Geographies* now places on the agenda in so eloquent and timely a fashion)" (1989, page 45). Soja refers to Harvey: "A brilliant example of this flexible halfway house of Late Modern Marxist geography is Harvey's recent paper..." (page 73) and Harvey is duly quoted on the back of Soja's book "One of the most challenging and stimulating books ever written on the thorny issue...." On the back of Harvey's book we have Soja: "Few people have penetrated the heartland of contemporary cultural theory and critique as explosively or as insightfully as David Harvey."[57]

These three themes, the situatedness of the author, the work as itself an element in a social system, and the limitations of representation (and the consequences for research subjects), were engaged in an angry and compelling way in the works I have just mentioned. But they need, too, to be seen as exemplars of a broader set of theoretical works within geography, all of which have attempted to develop a feminist theory of the written work. Drawing on a range of works in feminist theory more generally, these works have attempted to make sense of the written work against the background of that spectrum. In doing so they have addressed themselves to those central issues, such as the nature and limits of relativism, that have dogged postmodernists. In large measure, and because of their attention to the three central themes just mentioned, they have been able to articulate an alternative that at once serves the specific interests of those doing explicitly feminist scholarship and provides a challenge to others.

Learning from Feminist Theory

Perhaps most fundamentally, the advantage that feminist theories have had over the new common sense, as found in the new cultural geography, is this: that the formulation of the initial question within the new common sense has given the game away. It has led to positions from which its practitioners have found it impossible to extricate themselves. As I have suggested earlier, it has done this insofar as it has been led into a position wherein it is difficult to judge one position better than another. Indeed, it has done this because it has taken over from traditional common sense a set of views about the nature of science. Primary here has been the notion of science as consisting of either a universal or a nonlinguistic text.

We recall that in the modern era the work of science became increasingly disembodied. "Housed" within written works that appeared to be containers of ideas, science came to be seen as consisting in the first instance of theories and concepts that existed in a kind of Platonic netherworld. The scientist, at the same time, came to be seen as placeless, as engaging in work that was in no fundamental way situated but rather could be judged in terms that were believed to be utterly indifferent with respect to the scientist.

Now, when we turn to the new common sense, it may appear that we have moved away from this view. And indeed, the works of experimental ethnographers and anthropologists—like Vincent Crapanzano's *Tuhami: Portrait of a Moroccan,* Kevin Dwyer's *Moroccan Dialogues,* Marjorie Shostak's *Nisa: The Life and Words of a !Kung Woman*[58]—must strike one as far removed indeed from the works of Newton and Galileo. But in fact what we see is similar, for the following reason. In the new common sense the use of experimental textual forms is offered as a means for the solution of a problem that is itself theoretical in nature. The anthropologist begins by overturning a (presumed) previous view within which he or she was in a dominant position with respect to the research subject; from that it is imagined that a new and more adequate theory of that relationship arises, and from there one is led into the process of textual experimentation. But note that here, as in the traditional common sense, the point of departure is a set of concepts that are believed to be, or at least treated as though they are, outside of both the anthropologist and the subject. The anthropologist has begun by stipulating a theory, adopted on the basis of some set of criteria, and then attempted to apply it rather generally to whomever and whatever comes along. Granted, there are differences; for in the traditional common sense the criteria used for theory choice were associated with matters like "correspondence to reality," whereas in contemporary anthropology we have heard more about "the recognition of the fundamental humanness of our subjects," or "the rights of subjects to be heard." Yet in the end this is not a difference that makes a difference, for in both cases the starting point has been not a question but a stipulation. And because that stipulation occurs in the form of a mental act, where "the recognition of fundamental humanness" or "the rights of subjects" are viewed as nonlinguistic principles, the entire edifice associated with the production of knowledge comes to be devalued, comes to be seen merely as a sometimes shabby piece of scaffolding, useful but inessential.

In fact, it is because of the failings of this starting point that feminist scholarship, which begins elsewhere, has been able to provide a more adequate groundwork for the understanding of the written work. Put most simply, the starting point has not been (or at least has not been in the cases at hand) the stipulation of a theory of "the nature of women," but rather a set of questions, about the place of women in society, the extent to which that position is undesirable, and the ways in which it might be changed. Writing of anthropology, Frances Mascia-Lees, Patricia Sharpe, and Colleen Ballerino Cohen suggest that

> practitioners seeking to write a genuinely new ethnography would do better to use feminist theory as a model than to draw on postmodern trends in epistemology and literary criticism with which they have thus far claimed allegiance . . . [because it embodies] a politics skeptical and critical of "universal truths" concerning human behavior.[59]

And within geography, Susan Christopherson elaborates:

> [A] feminist would look at the problem of reconstructing human geography by asking political questions about "the project." Who would participate in defining this project? What questions are being subordinated to others? Why? Which daily life experiences constitute the material for theory? How are these experiences represented?[60]

Or as Linda McDowell put it, in feminist research,

> women's experiences, ideas and needs become accepted as valid in their own right. . . . [I]t is research *for* women. It should be useful, instrumental research, research that aims to make a contribution to improving women's lives in one way or another, research that contributes to liberation.[61]

If the beginning point is the asking of a set of questions, some of them quite practical in import, a necessary element of both the questions and the answers is the articulation of the situatedness of those being studied and those doing the studying. This articulation has turned out to be an area of extraordinary contention, in large measure because so much hinges upon the ways in which one characterizes this situatedness, the scale at which it is seen to be operating, and its durability. Jane Flax, for example, has seen three main positions, of feminist rationalism, feminine antirationalism, and feminist postrationalism.[62] And in a related analysis, Sandra Harding has distinguished feminist empiricism, feminist standpoint theory, and feminist postmodernism.[63] Seeing feminist empiricists as having too completely adopted the traditional strategies of modernist science, she describes standpoint theory as asserting that "the experiences arising from the activities assigned to women, understood through feminist theory, provide a starting point for developing potentially more complete and less distorted knowledge claims than do men's experiences."[64] One element of this view is the belief that one needs to begin one's analysis of the situatedness of women by seeing "the permeation of science as an institution and a system of thought by political life."[65] Harding says:

> [M]y daily actual activities, structured by social divisions of activity by gender, set limits on what I (and therefore, my culture) can see. Movements of social liberation make possible new kinds of human activity, and it is on the basis of this activity that new sciences can emerge.[66]

This is the crucial methodological point:

> The oppressed are indeed damaged by their social experience, but what is a disadvantage in terms of their oppression *can* become an advantage in terms of science: Starting off from administrative/managerial activity in

> order to explain the world insures more partial and distorted understandings than starting off from the contradictory activities of women scientists.[67]

So it was the very narrowness of the context within which modernist scientists were socially situated that led to the development of a science that was itself narrow. Within that science the elements outside the everyday experience of scientists, including much of women's lives, came to be characterized in extremely limited ways. According to Harding—and this turns out to be a central issue in feminist theory as it was in postmodernism—this narrowness of vision is the source, the cause of the development of theories that argue for some "essential" nature of women.

Of course, not everyone has agreed with her analysis; Jane Flax, for example, has stepped firmly into the postmodernist camp:

> Indeed, the notion of a feminist standpoint that is truer than previous (male) ones seems to rest upon many problematic and unexamined assumptions....
>
> I believe, on the contrary, that there is no force or reality outside our social relations and activity (e.g., history, reason, progress, science, some transcendental essence) that will rescue us from partiality and differences....
>
> If we do our work well, reality will appear even more unstable, complex, and disorderly than it does now.[68]

Here, though, we see once again the limitations of the new common sense. Where Harding has hinted at the ways in which certain social situations can be seen as providing the foundations on which essentialism is developed, for postmodernists and for advocates of the new common sense, essentialism is simply a bad habit, one that decent people avoid.

By contrast, I shall argue that here Harding is absolutely correct. But moreover, I shall argue that postmodernists have in setting their sights on foundationalism fundamentally missed the point. For a more adequate understanding of the nature of human practices, and of their relationship to images, will show that "foundationalism" is no more than an image, albeit one that is continually sustained by a wide range of social and technical factors—including the written work. So in our society in order to understand the nature of foundationalism (or related issues such as essentialism or relativism), it will be necessary to understand the nature of the written work, and in order to understand the nature of the written work, we need to understand something more about those concepts. Further, and finally, to understand more clearly the nature of image and practice, we must at the same time develop an adequate conceptualization of space and place. Indeed, as I shall argue in the next chapter, the two failures, to understand image and practice and to understand space and place, are intimately connected; it is this twin failure that diverts attention to issues like relativism, foundationalism, and essentialism.

CHAPTER FOUR

Beyond the New Common Sense: Toward a Geography of the Work in the World

Just as particular theoretical positions with respect to foundationalism (or relativism or essentialism) arise out of certain kinds of social situations, so do particular positions with respect to the written work. Here, of course, a great deal of interesting and useful work, especially on rhetoric and the sociology of science, has been done. But there has been something about the social situations from which that work has been written that has rendered it difficult for its authors to develop a broad understanding of the written work. I would argue that a fundamental source of that difficulty has been that its authors have been writing from a position that promotes a particular view of the nature of space and place, and the relationship between the two. And it is this view that reverberates through accounts of the written work, again and again stopping inquiry at fruitful junctures.

In what ways does it do so? First, it affects the understanding of the nature of the written work itself, in relation to the world. Second, it affects the understanding of the relation between the author and the work. And third, it affects the way in which common sense is transformed into what is seen as a broader or deeper theoretical understanding.

The issues of space and place are relevant to each of these effects, and in two ways. First, each is expressive of a deep-seated tendency toward the elision of the boundary between space and place, where it comes to be assumed that the two are fundamentally related, and place is seen as in some way derivative of space. Second, each is expressive of a persistent blindness toward certain views of the nature of space.

In what follows I shall first lay out the traditional ways of understanding the nature of space. I shall show that one finds a recurrent set of themes, and that far from involving a progressive rejection of incorrect or outmoded views and their replacement by new and more adequate ones, the history of the appeal to conceptions of space is a matter of the simultaneous appeal to a variety of those conceptions. Further, I shall suggest that there exist a number of reasons for the maintenance of one view over another within any particular arena; here explanatory power and consistency often take a back seat to other factors.

Second, I shall turn to the issue of place. I shall show that far from being merely a subset of space, places operate within a fundamentally different form of discourse. That is, the construction of places operates through a set of means very different from those that are operative in the discourse about space. Indeed, it will turn out that whereas in everyday discourse explicit and implicit appeal is regularly made to place and places, the appeal to space is made only under special circumstances.

Third, I shall turn briefly away from the issues of space and place, and note the ways in which the ordinary understanding of the relationship between objects and the space within which they are said to exist is a particular example of a broader understanding of the relationship between rule and application or theory and practice, an understanding that has the unfortunate effect of rendering it difficult indeed to recognize the ways in which everyday practices, including those associated with the written work, are grounded within particular places or specific kinds of places. By contrast, once we begin to see that in the matter of space what is often taken to be a matter of *theory* and application is more often a matter of *image* and practice, we can begin to see the ways in which particular spatial images, and especially the images of Newtonian and Kantian space, dominate the codification of the traditional common sense and the new common sense, in the ways in which they appear to offer theories of how things *must* be and how practice *must* proceed.

Finally, I shall suggest that once we recognize the ways in which the misapprehension of the nature of theory and application and of space and place color our understanding of the written work, we can begin to develop a broader understanding of that work. This understanding will help make clearer several vexing questions. Why is it so difficult to change one's mind? Why is it so difficult to offer views that are truly radical? And why are works so often "misunderstood"?

On Conceptions of Space

In Western thought there have really been only four main notions of space. Each has gained wide popularity, but each has at the same time been formally codified by a scientist or philosopher. I shall here refer to those notions in terms of the formal codifications, but it is important to see that each has had an existence apart from the work of individual scientists or philosophers. The first, codified by Aristotle, is static, hierarchical, and concrete. It gives greatest attention to a concept of place. The second, which we usually associate with Newton, imagines space as a kind of absolute grid, within which objects are located and events occur. The third, found in Leibniz, adopts the scientific outlook of Newton but argues that we need, as in Aristotle, to attend to the relationships among objects and events, to the extent that we come to see space as fundamentally relational and defined entirely in terms of those relationships. The last, codified by Kant, turns the tables; whereas Aristotle and Newton had seen discussions of space as essentially about the world, Kant argued that we need to see space as a form imposed on the world by humans.

Each of these notions constitutes a powerful image of space. But of the three, the Newtonian is the one both most familiar and most often imagined to be accurate and to govern our activities and thoughts. Most people imagine that after all is said and done, they live in a Newtonian world. I shall, though, deny that this is true. In fact, I shall argue that of the three, the Aristotelian is by far the most important in everyday life, and in the practice of geography. Moreover, the Leibnizian, although little noticed, is fundamental to thinking within geography and the social sciences. The Newtonian, I shall grant, is also of fundamental importance but largely because it is such a powerful *image*, and an image supported in so many ways. And the Kantian, despite its role in cultural studies and its popularity today in the form of the neo-Kantian view that all knowledge is somehow relative to the position of the speaker, exists only against the looming presence of a Newtonian absolutism.

ON ARISTOTLE AND THE NATURAL PLACE

When we think of the influence of Greek thinkers on geography, we normally think first of two people, Eratosthenes and Strabo.[1] Eratosthenes is sometimes characterized as the "father of geography," in part because he was the first person to use the term and in part because of his famous attempt to measure the circumference of the earth and thereby to take a mathematical and hence scientific view of it. George Sarton, for example, has called him—on the basis of this attempt—"one of the greatest geographers of all ages."[2] And Strabo is widely remembered as the person whose monumental work summarized for the future the geographical knowledge of the time. But although both Eratosthenes and Strabo have been important, it is difficult, really, to see their works as essential to, or even connected with, the current practice of geography. Indeed, characterizations like that of Fred Lukermann, to the extent that they show Strabo as developing a view of geography that involved a rich understanding of the interrelatedness of geography, chorography, and topography, simply demonstrate the extent to which modern-day geographers have failed to appreciate what is important in the work of their predecessors.[3] Would contemporary geography be any different if they had never existed? It is hard to imagine that it would.

By contrast, the views expressed in and first codified by Aristotle remain of overwhelming importance today.[4] It would be easy here to misunderstand me. I do not mean to suggest that he was a person of such genius that he invented a philosophical and scientific system that just through the force of that genius has survived. Rather, I suggest that he was the first writer to notice and elucidate something about the Western way of inhabiting and relating to space. His view has persisted for two reasons: the first is that what he noticed was then so important, and the second is that his codification of it—because it was so popular—became deeply ingrained in discourse not only about space and place but about the world more generally. Long after his lifetime his view shaped Western discourse about space. When it comes to matters of space, we speak Aristotelian.

Now, Aristotle's physics is based on the belief that what needs to be explained is change, and particularly motion, change in location. Here he assumed that there needed to be a reason or cause for any motion, that motion did not simply occur. But to experience there are really two sorts of motion. On the one hand, there is the motion of the planets and stars, and that motion by all accounts was circular and eternal; at least in his time and given the instruments available, there was no evidence of changes in such motions. On the other hand, there was the sort of change that we see on the earth, as when I throw a rock in the air and it falls to earth. To Aristotle and many of his contemporaries, it seemed as though we actually needed two very different sets of explanations for those two apparently different sets of phenomena. In the celestial realm it appeared as though it was circular motion that needed explanation; in the terrestrial, it was linear.

The explanation for the terrestrial is the more important here, and it drew upon the belief that the world was composed of four substances: earth, water, air, and fire. It did not escape Aristotle's notice that the earth appeared by and large to be composed of earth, that the oceans lay upon the earth and were composed of water, that next there was the air, and finally fire. In fact, when left unfettered, earth appeared naturally to fall to earth, water to fall to water, air to rise to air, and fire to continue upward. This change in location, he believed, was what characterized the terrestrial sphere, where what we see is constant degeneration and regeneration. And it is precisely here where the notion of place becomes important because for Aristotle things tend naturally to move toward their own places; indeed, this is the very nature of natural motion.

By contrast, when an object is moved away from its natural place, as when a rock is thrown into the air, what we see is not natural, but "violent" motion. When that motion ceases, when its motive force is removed, the rock tends to return to its own natural location. This natural motion tends toward the natural place of things, that is, to their natural sphere. Because earth is the heaviest of the elements, things made of earth tend toward the center of the universe, which is the center of the earth. It is important here to see that for Aristotle objects made of earth fall to the center because it is the center, and not because they somehow are attracted by a gravitylike force to the mass already there; indeed, were the earth located somewhere else, objects would still fall toward the natural center.

A final feature of his physics is also of importance. It is that although Aristotle believed that there is a relationship between the weight of something and the way in which it tends to fall, so that heavier objects fall faster than light ones, there are only a few places—his discussion of the circumference of the earth is a notable one—where he attempts to quantify phenomena. His work is overwhelmingly qualitative in thrust. It might be thought that his physics, because it was a qualitative science based on the belief in the natural place of things, has long since been superseded, replaced by something that makes more sense. But before leaping to that conclusion, it is important to note that what many today take to be the commonsense view of things, that the universe is a void full of atoms, and that space is a featureless grid, was well known to Aristotle. That very view had been developed in some detail by the Greek atomists and was eloquently laid out, after Aristotle's time, in Lucretius's *On the Nature of Things.* So Aristotle's failure to adopt this view was not a matter of its not having occurred to him. Rather, it was a result of his belief that such a view was simply incapable of being developed in a way that saved the phenomena, that accounted for what was obvious to experience. Indeed, current historical work on the development of alternatives to Aristotle's view suggests that far from being attempts to develop theories that better fit the phenomena, they often were less, rather than more, successful than their predecessors. Their appeal lay elsewhere.

But what does it mean to say that geographers, and people more generally, live today in an Aristotelian world? There is, of course, the obvious answer: notwithstanding the Copernican revolution, everyday experience still tells us that the earth stands still while the sun, moon, planets, and stars move through the sky in a circular motion. Further, although we are now aware that stars come and go, the heavens appear to change little, and certainly little in comparison with what we see every day on the earth; for most of us the evidence of supernovas comes from the mass media and not from our own experiences. Moreover, the evidence of our senses suggests the truth of the view that water returns to water, earth to earth. Few of us can say that we have experienced the pull of gravity, but most have seen rivers, landslides, or the rise of a bubble through a body of water. Last, and perhaps most important, we live in a hierarchical world where things and people have places where they belong. We are taught early what it means to be out of place.

THE COLLAPSE OF HIERARCHY

Aristotle's view was enormously influential, and because of its adoption by the Roman Catholic Church, remained so well through the Middle Ages.[5] But in the seventeenth century it was displaced by a set of views developed by a group, including Descartes, Boyle, Galileo, Newton, and Leibniz, whose works came to establish a radically different understanding of space. Many would argue that these views have totally replaced the earlier view. And certainly, when we speak today of space, we likely think immediately of Descartes, Newton, or, for some, Leibniz. But it should be clear from what I have just said that I believe this not to be the case.

Rather, what has prevailed is a set of *images* of space, which guide the ways in which people think about space but do not necessarily affect the ways in which they actually organize or act within space. This new image, I would suggest, has prevailed for two somewhat different reasons. First, it fit well into a technological consciousness that emerged as early as Roman land surveying and military organization, and that flourished alongside these conceptions of space.[6] And second, and in a sense more important, it provided an image of clarity of thought, a vision that is of such power that in a wide variety of areas, in art, architecture, and politics as well as science and engineering, it came to be seen as defining the modern age. Its power has been such that it has been able to render almost invisible the omnipresent remnants of the Aristotelian view.

Actually, and as I suggested earlier, we need to distinguish here between two different attempts to develop mechanical images of space, one found in Newton and Descartes, the other developed by Leibniz. Newton in the end won out, but not before standing, through an intermediary, in battle with Leibniz. Leibniz's view, though, remains important, both because it provided an alternative to what was to become the orthodox image and because geographers today appeal to it in so many ways.

In his *Principles of Philosophy* Descartes argues that "a space, or intrinsic place, does not differ in actuality from the body that occupies it; the difference lies simply in our ordinary ways of thinking. . . ."[7] And he says:

> The terms *place* and *space* do not signify something different from the body that is said to be in a place; they merely mean its size, shape, and position relative to other bodies. To determine the position we have to look to some other bodies, regarded as unmoving.[8]

So for him there can be no void, no empty space. Rather, we need to begin with the understanding that all of the characteristics, color, density, and so on, that we associate with objects in space are really inessential features, and that the only essential feature of objects is their extension, their length, breadth, and height. But once we remove those inessential features, we then see that what is left, extension, is just the same as space. Indeed, space cannot be seen as existing without matter and extension, so space consists simply of the relations among extended objects.

Even so, Descartes believed that space was without limit:

> We see, furthermore, that this world—the totality of corporeal substance—has no limits to its extension. Wherever we imagine the boundaries to be, there is always the possibility, not merely of imagining further space indefinitely extended, but also of seeing that this imagination is true to fact—that such space actually exists.[9]

And from this it follows that where for Aristotle there had been one physics for the earth and another for the area outside of the moon, for Descartes there is only one:

> We can also readily derive the result that celestial and terrestrial matter do not differ; if these were an infinity of worlds, they could not but consist of one and the same kind of matter; and thus there cannot be a plurality of worlds, but only one. . . .
>
> Thus it is one and the same matter that exists throughout the universe. . . . [10]

Although this view is in some respects familiar, it is in others quite puzzling, and especially in its identification of space and matter. And it is one that has more than a few shades of Aristotle. Yet behind it we need to see three related features, which were to overwhelm the Aristotelian elements of that work and thereby point to a more modern view. First, in his *Geometry* Descartes developed the connection between algebra and geometry; this made it possible to move beyond the traditional picture of mathematics. Second, he presented a view wherein mathematics is the model of certainty, and indeed of all knowledge. Aristotle's physics had been purely qualitative, and in the Middle Ages the church had actually proscribed the application of mathematics to science; now mathematics became not merely a tool for science but the very model of science. And finally, as we saw in chapter 1, he developed a view wherein knowledge developed as a result of "mental vision." Here, echoing Brunelleschi's system of linear perspective, he laid out a way of thinking about the act of acquiring knowledge that made it possible to see the knower as standing outside of any possible situation and viewing it from a detached position.

NEWTON AND ABSOLUTE SPACE

The implications of these views were worked out by Newton in his 1686 *Mathematical Principles of Natural Philosophy.*[11] In the famous Scholium to the Definitions, he laid them out in the starkest and most straightforward way:

> Absolute space, in its own nature, without relation to anything external, remains always similar and immovable. Relative space is some movable dimension or measure of the absolute spaces. . . . Absolute and relative space are the same figure and magnitude, but they do not always remain numerically the same. For if the earth, for instance, moves, a space of our air, which relatively and in respect of the earth remains always the same, will at one time be one part of the absolute space into which the air passes; at another time it will be another part of the same, and so, absolutely understood, it will be continually changed.[12]

One consequence is that we need to think of motion in a very different way:

> Absolute motion is the translation of a body from one absolute place into another, and relative motion the translation from one relative place into another. Thus in a ship under sail the relative place of a body is that part of the ship which the body possesses. . . . Wherefore, if the earth is really at rest, the body, which relatively rests in the ship, will really and absolutely move with the same velocity which the ship has on earth.[13]

Why do people fail to see that space is absolute?

> [B]ecause the parts of space cannot be seen or distinguished from one another by our senses, therefore in their stead we use sensible measures of them. . . . And so, instead of absolute places and motions, we use relative ones, . . . but in philosophical disquisitions, we ought to abstract from our senses and consider things themselves, distinct from what are only sensible measures of them.[14]

Whereas today we may see Newton's work as strictly secular, we need to recall that it in fact has substantial religious overtones. In his General Scholium he argued that God

> is eternal and infinite, omnipotent and omniscient. . . . He endures forever and is everywhere present. He is omnipresent not *virtually* only but also *substantially*. . . . In him are all things contained and moved, yet neither affects the other; God suffers nothing from the motion of bodies, bodies find no resistance from the omnipresence of God.[15]

Yet those overtones have long since faded away, and what has remained is a view of a universe of matter, floating in a space that is infinite, absolute, and eternal. This is an image of extraordinary power. It is an image that seems strikingly clear, strikingly subject to quantification, and surely consistent with the requirements of science. Notwithstanding the twentieth century's flirtations with alternatives, via the popularizations of Einstein's work on space and time, this image remains a basic feature of common sense.

LEIBNIZ AND RELATIONAL SPACE

But not everyone has been seduced by the power of Newton's view. Indeed, Leibniz, Newton's contemporary, lashed out, arguing that Newton's view of space was literally nonsensical; in a series of letters with Newton's proxy, Samuel Clarke, he developed this view.[16] In a famous passage he noted how people come to believe in space through the concept of motion:

> They consider that many things exist at once and they observe in them a certain order of co-existence, according to which the relation of one thing to another is more or less simple.... When it happens that one of those co-existent things changes its relation to a multitude of others, which do not change their relation among themselves we then say, it is come into the place of the former; and this change we call a motion in that body.[17]

The way in which we understand motion leads to the belief in the existence of absolute space:

> And supposing, or feigning, that among those co-existents, there is a sufficient number of them, which have undergone no change; then we may say, that those which have such a relation to those fixed existents, as others had to them before, have now the *same place* which those others had. And that which comprehends all those places, is called *space*.[18]

So the belief in the existence of absolute space arises simply by virtue of our noticing that among the objects in the world, some appear to move, whereas some appear not to. Yet to go along with Newton and move from this perception to the conclusion there is something called "absolute space" is to move from the realm of science to that of metaphysics.[19] In order to stay within the bounds of science, he argued, we need to understand that space is nothing more than "something merely relative, as time is; that I hold it to be an order of coexistences, as time is an order of successions."[20] Space, that is—and as in Descartes—is purely relational. But in contrast to Descartes, Leibniz does not believe that space and matter are identical; space is relational but consists just in those relations, and nothing else.

KANT AND THE SECOND COPERNICAN REVOLUTION

The dispute between Newton (or his proxy, Clarke) and Leibniz took place between 1715 and 1717. And for the next sixty-five years they seemed to have laid out the main alternatives in discussions about the nature of space. But in 1781 Kant published the *Critique of Pure Reason*, a work that fundamentally recast debates about the nature of space.[21] His view, after those of Aristotle, Newton, and Leibniz, is the final of the four alternatives widely subscribed to today.

Kant described himself as having created a "Copernican revolution" in philosophy. Where Copernicus had moved the earth from the center of the universe and replaced it with the sun, Kant had moved the locus of debate about knowledge from the known to the knower. In doing so he recast the question about the nature of space from one about the nature of the world to one about the nature of human knowledge.

His work is notoriously difficult, but simplifying greatly, he argued that previous accounts of the nature of knowledge, and here he referred to scientific knowledge, could not make sense of that knowledge. In particular, we know that there are certain branches of knowledge, like mathematics, where we can have certain knowledge, but we never really have a perception of certainty, we never actually see or experience it in the world. From where, then, does it come? His Copernican revolution claimed that the certainty comes from within, that it is built in to the way in which we know the world. So, for example, we never actually perceive causes in the world, but we naturally impute causality to the relationships among objects.

If this is true of the imputation of conceptual relationships like causality, it is also true more basically of space and time. In fact, we never actually perceive either. Rather we perceive a series of instants, or we perceive objects close to or far from one from another, or objects that seem to occupy volume. But, Kant argues in the case of space, if we did not already have built into us in some way the notion of space, the possibility of ordering things in spatial terms, we would be unable to say "This is next to this." As he put it,

> Space is not an empirical concept which has been derived from outer experiences. For in order that certain sensations be referred to something outside me (that is, to something in another region of space from that in which I find myself), and similarly in order that I may be able to represent them as outside and alongside one another, and accordingly as not only different but in different places, the representation of space must be presupposed.[22]

Thus, the whole way of discussing space that we find in Aristotle, Newton, and Leibniz needs to be abandoned. It is not only wrong but incoherent. For if space is a condition of our understanding the outside world, we can never ask "What is the world really like spatially?" All of our perceptions of the world are already spatial; we will never be able to get beyond our own perceptions.[23]

Kant himself believed strongly in the truth of Euclid's account of geometry and Newton's account of the physical world; indeed, he believed both to be final and definitive. And yet, by recasting the question of the nature of knowledge in this way, he opened the door for a critique of Newton and Euclid and for the development in the nineteenth and twentieth centuries of more complex understandings of the nature of space. In fact, it would not be going too far to say that all studies of culture are today a footnote to Kant. Exactly how this happened is not relevant here, but what is, is this:

during those two centuries Euclidean geometry, clock time, and Newtonian physics each came to be seen as only a single of several possible alternatives. The development of non-Euclidean geometry, for example, suggested that we may perceive space in radically variable ways.

The undercutting of the belief in the universality of traditional absolutist views of space was augmented both by the increasing awareness of alternative cultures and by the romantic reaction, in the West, to the increasing force of industrialism and urbanism. And so, by the turn of the twentieth century and by the time that geography began to be constituted as a modern discipline, a set of views that would allow for the belief in real differences in the human occupation of the world and organization of space were solidly in place. But so too were the Newtonian view of space as absolute and the Leibnizian view of space as relational.

On Places

It is common enough to believe that space constitutes a kind of general medium, and that places merely exist within that medium, as locations. Admitting that it is common to view places as having emotional or other cultural characteristics, some would say that a place is a "location plus meaning." But this view seems to me to be quite wrong, in large measure because it fails to see that the constitution of places and the development of conceptions of space are connected, but not connected in a simple and straightforward way. We shall see in the remaining sections of this chapter that there are good reasons why this relationship is not straightforward, but it will be helpful before turning to that issue to look more closely at the nature of places. Among a number of reasons for attending to the issue of the nature of places, certainly foremost here is that the written work always exists only in the context of a set of places appropriate to its use.

When I speak of a place, I use the term as conceptualized by a long series of geographers, extending in some respects back to Carl Sauer and Paul Vidal de la Blache but deriving more recently from the work of Yi-Fu Tuan.[24] The premise is that people inhabit a world that consists of places, that "place" is a fundamental feature of human experience. By place I mean quite literally that. Kansas is a place, and so is Los Angeles and Mt. McKinley and my office.

To say something is a place is not to say that it has some natural boundaries, ones that somehow existed long before people were there, but rather that it is a location that has been given shape and form by people. Although this is obvious in the case of Kansas or Los Angeles, it is only less obvious, but not less true, in the case of Mt. McKinley, because Mt. McKinley would not truly be a place if no one had ever visited it, if no one had named it, if no one had decided where it ended and the next peak began. As Sauer said, there are no natural places.[25] Indeed, Mt. McKinley is an especially apt example because officially it no longer exists; it has been replaced by "Denali," a name that embeds the place in a very different set of narratives and practices.

In fact, there are a number of ways in which people create places. We do so by naming them. The first thing that settlers do when they come to a place is to give it a name; they use that name to carve out a portion of what was inchoate and turn it into a place, of which they can talk and to which they can return. Here one can turn for more detail to Paul Carter's *The Road to Botany Bay,* a detailed account of the fundamental way in which naming was at the root of the construction of what is now Australia.[26] And one can turn, too, to the many governmental bodies charged with tracking geographical names and resolving problems associated with them.

A second way in which people create places also uses language; we create places by applying typologies. This is a river, this a stream. This is a ghetto, this is a suburb, or an inner city. We make places by coming to see what is new to us as a case of what is familiar. On this subject there is a large literature, detailing, for example, the ways in which the categorization of an area as a "desert" can be central to its existence.[27]

A third way of making a place is by making—or picking out—a symbol. Yosemite is a symbol of the West, the Statue of Liberty of America, the pyramids of Egypt. Here the part stands for the whole, and as anyone who has tried to argue in favor of burning the American flag knows, many people take the part to be the whole.

A fourth way of making places is by telling stories. In America we for many years told stories of the adventurous Columbus and of the heroic military fighting off the Indians; these, and stories in which good seemed always to win over evil, were ways in which we described ourselves as Americans, in which we made the nation into a place. And the failure of these older stories to resonate with new generations of women, African Americans, and immigrants is surely a significant part of the reason why those who believed in the older stories have come lately to feel the country to be disintegrating.

Finally, we make places by doing things. Some of these practices or habits are, in fact, highly ritualized. For example, in the settlement of the Western

Hemisphere by Europeans, there were prescribed rituals that declared ownership. When Columbus claimed Hispanola, for example, he made a public proclamation and unfurled banners. This proclamation was typically recorded by a notary and witnessed by as many people as possible. But a second step was necessary. An instruction given to De Solís in 1514 required

> cutting trees and boughs, and digging or making, if there be an opportunity, some small building, which should be in a part where there is some marked hill or a large tree, and you shall say how many leagues it is from the sea, a little more or less, and in which part, and what signs it has, and you shall make a gallows there, and have somebody bring a complaint before you, and as our captain and judge you shall pronounce upon and determine it, so that, in all, you shall take said possession. . . . [28]

Cortez, it is said, "moved walking on the said land from one part to another, and throwing sand from one part to another, and with his sword he struck certain trees . . . and did other acts of possession."[29]

But some of these rituals are far less formalized; when I move to a new town, I stake it out as my own place by finding a place to live, a route to work, the local grocery, the Irish pub, and so on and so on, by establishing a set of routines along familiar routes that come finally to be associated with names and typologies, symbols and stories.

Tuan has argued that we need to distinguish between people having a "sense of place" and being "rooted," and this turns out to be an important distinction.[30] He makes this distinction in a discussion of historic preservation. On this view, when the issue of preservation arises, people come consciously to consider just what it is about a place that makes it important and that, indeed, makes it what it is. Yet this having a "sense of place" is very different from the relationship that some people, and especially people in less developed societies, have with the places in which they live. In those places the relationship with a place is almost entirely one developed through engaging in practices (including the practice of telling stories), and there is little of what we term a "sense of place."

Although this strikes me as an important distinction, it seems to me that the process of being unselfconsciously rooted in a place is far more prevalent and fundamental than Tuan allows in his discussion of rootedness. Indeed, much of what goes on in everyday life is a matter of engaging in sets of practices that are only sometimes, and briefly, reflected upon; much of everyday life, whether in the car or in front of a computer or in a library, is very much a matter of rootedness, that is, a matter of everyday habit and routine.

The Confusion of Space and Place

What makes it difficult to see the importance of places is the propensity to confuse place and space. It may appear something of a shock for me to say that the two are in fundamental respects different, yet it is both true and important—and fundamentally relevant to the understanding of the nature of the written work. It is important because the ordinary understanding of the relationship between space and place is a particular example of a broader understanding of the relationship between rule and application or theory and practice. In both cases the particular is seen as an element of the general, and as an element fundamentally contained within that more general structure. So one typically sees places as elements within space, just as one sees the application of a rule as built into the rule. And to take this analogy further, it is common in science to imagine a theory as a spatial structure, which maps onto its facts in the same way that the points contained in a map of a region of space point to the places mapped. That is, the practice of science very often itself operates through the appeal to an image within which space and place are confused.

Although this appeal to these images appears to be very much commonsensical, when we begin looking at particular cases, of rules and applications or space and place, difficulties immediately arise. Perhaps the best way of understanding the nature of these difficulties is to consider the relationship between space and place, and then to turn to the issue of rule and application.

If we turn to the Newtonian image of space, we are likely inclined to imagine that his image, of the world as an object located in a vast and undifferentiated space, is the one that we all use, to which we all appeal. The facts, though, are quite otherwise. It may seem baffling that I say this. But if we turn back to the debate between Leibniz and Newton, we can see why. Recall that Newton argued in favor of the independent and absolute reality of space, and that Leibniz countered that the whole notion was incoherent because we could never tell where we really were. Newton's argument, like the arguments of the other absolutists (as in Kant's early work on space) was *not* that we have evidence that space is absolute and independent, but rather that we *must* believe that space is independent and absolute. Leibniz's argument is that space is relational because our only way of determining where things are appeals to relations; Newton makes a wholly different argument, that if we do not believe in absolute space, we must as a consequence abandon other beliefs, and that those other beliefs are not ones we are prepared to abandon.

In fact, then, Newton presents less a falsifiable *theory* of space than an *image* of space. And in geography, as elsewhere, it is as an image that it has been important. The image, in effect, provides a backdrop against which those who attempt to develop timeless and universal theories can set their work. In one sense this is a matter of the establishment of what look like grounds for believing that work can be permanent, that the researcher can come back later to what is truly the "same place." But there is another, and perhaps more important, sense in which the Newtonian image has been important, and that is as a model for the very operation of society. Newton's work was the very culmination of what E. J. Dijksterhuis has called the "mechanization of the world picture."[31] And as such, it provided an image of the ways in which things work, and so, an image of the kinds of explanations that ought to be offered. These explanations ought, it has been concluded, to see the world as composed of elements or actions that are interchangeable and that maintain their characteristics when moved to different places.

The influence of this view is obvious. In the classic works on central places, agricultural location, and industrial location, the assumption was always that there was some crucial element of the explanation, often utility or buying power, that was absolutely footloose. And we see the same view in other, perhaps less obvious, places. Early Marxist works, for example, denied the reality of place and nature but viewed use and exchange value as mobile elements; their critique of capitalism was that it treated the worker as an atom in Newtonian space, but this critique appealed in its own way to that same image. Finally, contemporary works on spatial perception have tended to adopt a view that is Newtonian in another way. They have couched their analyses of the spatial perceptions of others in terms of a kind of "deviation from the real," where the real is absolute and unchanging—and where the geographer is able for the purpose of analyses to step outside of space.

Hence, if the Newtonian view of space has in various ways provided a guiding image for those engaged in geographical work, its very nature—as a view supported by argument rather than evidence—has prevented its use as anything other than an image. Indeed, it is difficult to find a point at which the *practice* of geography could itself be said to be absolutist with respect to space. Rather, the concept of "absolute" space enters into geographical practice not as an overriding theoretical construct but rather just where the geographer finds it necessary or convenient to provide the audience with a tertium quid, with an image that will connect the author and the audience through something that they all believe themselves to understand and accept. And this belief is in turn undergirded both by a certain class of experiences and by the existence within society of related sets of images. The class of experiences consists of those very simple and basic ones that appeal to an object as being inside another; just as a fish is in a barrel, so too can an object or place be said to be "within" a spatial container.

At the same time, everyday experience, and certainly such experience in our society, appears at almost every turn to support the notion that space is absolute and nondirectional. We see this in floor tiles and light switches, in the wide range of experiences of a geometricized world. Perhaps most important, we see this appeal to such spatial imagery in our society in the dual images of Fordism and Taylorism, in the attempt through geometry to develop a more rational and efficient means of production.[32] Yet although these examples suggest that there is a real sense in which one can see a society being transfigured by the image of absolute space into one in which absolute space was indeed central, it is useful as a counterpoint to consider the example described by David Noble. In his provocative *Forces of Production,* Noble describes the "battle" between two different groups to establish the fundamental structure of numerical control on the factory floor.[33] One group, advocates of what might best be termed "analog" production, argued that the best way to reproduce certain objects was to attach a recorder to a traditional machine and then have an experienced machinist create a prototype. Subsequent copies could be created simply by playing back the original. By contrast, advocates of "digital" production wished to remove control from the unionized shop floor and place it in the hands of white-collar workers, who would create first plans and then digital images of the final product, which would be produced without the intervention of a human machinist.

At first glance, of course, it appears that Noble has described a case in the real world where one sees the machinist creating an object through a set of practices, and the white-collar worker creating the same object through appeal to a spatial, Newtonian construct. Yet on reflection it ought to be clear that what we are really seeing is rather a movement of practices back a step; now the creation of the object through digital control is only the last step of a process in which the white-collar worker has engaged in a set of practices as routine and habitual as those engaged in by the traditional machinist. The white-collar worker appeals in describing the production of an object to a Cartesian or Newtonian spatial structure, but that appeal is itself in the form of a practice, and not of an overriding theoretical construct.

Noble, concerned especially with the fortunes of the machinist, does not press this point; Shoshana Zuboff, however, does.[34] In her work *In the Age of the Smart Machine: The Future of Work and Power,* she shows the ways in which the introduction of computer technology into the workplace does not do away with the forms of bodily discipline traditionally associated with labor but rather recasts them. The practice of labor may be different, but in the most intellectually demanding activities it remains that, a practice.

If we turn now to a second notion of space, the Kantian, we find rather a different relationship between that conception of space and the practices involved in the construction of everyday places. As I suggested earlier, this view directs its attention in a very different way. Where the others saw the question of the nature of space as in the first instance a question about the nature of the world, for the Kantian the tables are turned, and it becomes a question about the nature of the observer. For Kant himself this view changed little, and Newton was still right, but for those who have followed in his footsteps, it has been an easy move to the belief that different people live in worlds that are themselves spatially very different.

Now, the Kantian turn in the understanding of space has appeared to be most clearly in evidence in the area of perception studies, where beginning in the 1970s a group of geographers began to consider the perception of space and hazards among various groups. Differential perceptions were, of course, central to the work in hazards.[35] But in the case of hazards research, where the interest was in the perception by residents of floodplains, for example, of the risks of living there, the aim was always to compare the perceived risks with the "real" risks. And to the extent that this notion of the "real" was not problematized by calling into question the objectivity of the researcher, this is not really a Kantian view at all, since it does rest on the belief that there is an identifiable real.

Much the same can be said of most of the work on spatial perception; although it appears on the face of it to be concerned with space-as-perceived, there is almost always the underlying assumption that perceived space can be compared with the space that is "really out there." There is, of course, another set of works on space and place, that done by people whose concern might best be termed "cultural." These works have tended not to be explicitly reflexive in form but have nonetheless made it clear that the means of conceptual or other ordering of space and place that they describe ought to be seen as applying as much to the author as to the subject.[36]

And so, if we leave our sights on the general conceptual form of geographical arguments, it turns out to be quite difficult to come up with examples of works that see *all* perceptions of the world as structured by the knowing subject; it is far more common to see it assumed that everyone but ourselves is seeing the world through conceptual blinkers, while we ourselves are the only ones able to see it as it really is.

At the same time, a sort of Kantianism does very often enter into certain characterizations of everyday experience, and in a way very much reminiscent of the way in which Newtonianism is invoked. Kantianism enters into the reflective explication of the actions of others when those others appear difficult to comprehend. Here the common response is to claim that the others "have a different way of looking at the world" or a different "worldview." It is important to notice that what is being said is *not* that the others act in odd or unusual ways; indeed, that is being presumed. What is being claimed is that in order to understand those ways of acting, we need to offer reflectively an explanation that appeals to something internal, something mental. Yet as in the case of the appeal to Newtonian conceptions of space, the appeal to Kantian notions looks like an appeal that can subsume particular actions under a general schema, but it can better be seen as an appeal to an image that offers to gather together seemingly disparate phenomena. Having stated a series of apparently incongruous facts about an individual or group, we tie them together by saying, "They have a different worldview."

There remain two conceptions of space, the Leibnizian relational one and the Aristotelian, that enter into characterizations of everyday practice in ways different from the Newtonian and the Kantian. Indeed, both are more directly involved in those everyday practices, more richly implicated in everyday life.

This statement may seem a striking and wishful one to anyone who has read Leibniz, and I do not mean to suggest that we go around talking about monads and possible worlds. Rather, we need to turn to far more prosaic matters. For example, if we turn back to Euclid, or then to Brunelleschi, we likely see their work, on geometry and then on linear perspective, as presaging the development of the Newtonian view. This, though, is quite wrong. Euclid, for example, attempted to show how certain truths could be deduced from a set of axioms and postulates. In effect, he argued that one could construct the system of geometry simply from them. What is missing here? The rest of the world. In fact, Euclid's geometrical system is completely closed; it exists simply *as* a system. Much the same can be said of linear perspective in painting. From our point of view it looks as though the Renaissance development of the system of perspective in painting was an attempt to make painting consistent with what later came to be the absolutist, Newtonian view. But in fact, as in Euclid, the system of the painting is quite closed; it includes only the elements of the painting, along with the viewer. Nothing else matters.

And if we turn to geography, and to works that look Newtonian, we find much the same. The most abstract models are inevitably simply that, models. They consist of elements and relations that are—and indeed, endeavor to be—utterly self-contained. Today this is perhaps most obvious in the case of geographic information systems; although they appear to be absolutist in intent, and although they certainly are in imagery, they are distinctively relational in their understanding of space. Where one moves beyond the model and theory, matters change very little. Descriptions of the movement of people and goods are couched in terms of locations, but the locations are never absolute; they are always, in turn, characterized in terms of yet other locations, in what ultimately becomes a self-enclosed system.

And finally, turning away from the grammar of everyday scientific practice to that of everyday discourse, it is common in that discourse to invoke the traditional geographic distinction between site and situation. This, of course, is the distinction between characteristics that can be said to inhere in a place, characteristics of vegetation or human occupancy, and characteristics said instead to be relational in Leibniz's sense, to express the relation between a particular location and its environs.

So in these cases of theory and practice, and certainly more than in the cases of the Newtonian and the Kantian, one finds far more clear-cut the reference to a conception of space that is fundamentally relational; that is, a basically Leibnizian framework is more readily invoked as a central element of the apprehension of the world and of the engagement in everyday practices than are the frameworks of Newton and Kant.

And much the same can be said of the final, Aristotelian view. For as foreign as Aristotle may seem, when his view is stripped to its bare lineaments, it must strike one as closely related indeed to the usual characterizations of the commonsense world. This may seem a rash assertion, because so few today would claim to have intellectual roots extending much beyond Newton. And yet, if we look at the geography of the everyday practice of geographers, it is shot through with Aristotelian assumptions. Most important is that we can see the world in terms of objects and events, each of which has its natural place. For Aristotle earth, air, fire, and water had a natural place; for the contemporary geographer the list is different; women, ethnic groups, economic activity, trees, and rocks all today have their own natural places. Indeed, to be out of place is to be a possible subject of research, and to be comfortably where one belongs is to be rendered invisible. And so, until very recently geographers assumed that women belonged in the home. As a result, they paid virtually no attention to them. Studies of economics and culture focused on men, and data categories were established in ways that made the activities in which women engaged difficult to see. It was only when women became visible on two fronts—by making it clear that the home was not the only place where they might fit and by taking up stronger positions outside the home—that geographers began to look more closely at the position of women in society and to notice that they had, all along, been engaged in important activity.

This example, it should be noted, points to an important feature of the contemporary version of Aristotelianism; that is, that to refer to something as being in its "natural" place is not simply to make a factual statement; it is also to make an evaluative claim. In contemporary society, to say that something is where it belongs is to say that it ought to be there. Nowhere is this more true than in the thinking of academic geographers about where they themselves belong. Even as those who characterize themselves as radical have criticized the academy, they have at the same time claimed—as have women and members of ethnic groups—that they *belong* there. So here, as in the case of the relational, one finds a conception of space invoked commonly in discourse about everyday life. Indeed, here it is the use of this conception to structure *theoretical* notions that appears to be missing.

Now, I have suggested that there are a variety of ways in which people construct a world of places out of some bare raw material. At the same time, at least for us in the West there are only a few (I have named four) central schemes used to conceptualize space. But although it would seem clear that one ought somehow to be able to derive a specific conception of place from each conception of space, things have turned out to be far more complex than that. A view of space often viewed as long outmoded, the Aristotelian, appears in fact to be right at the heart of everyday experience. One dauntingly abstract view, the Leibnizian, appears often to be invoked in both theory and practice, in the construction of theories and in the everyday appeal to site and situation. By contrast, the two most visible views, the Newtonian and the Kantian, tend to be invoked more as images, as post hoc means of supplying order to disparate worlds, either by appealing to the intrinsic order of the external world or by appealing to the knowing subject as a source of that order.

In pointing out these features of contemporary geographic practice, I am not, I hasten to add, attempting to be critical of those who appeal to one or another of these terms. (At least, I am not criticizing them insofar as they are ready to notice that that is what they are doing.) Quite to the contrary, it appears to me that these sorts of appeals are quite inescapable, to the extent that they turn out to be institutionalized into everyday life. And it further seems to me that it is this fact that both renders the written work less in the control of the author and makes it difficult for the author to pull off a true change of heart. Certain of those ways of institutionalization of the written work will be my subject in the remaining three chapters of this volume.

Image and Practice

But before turning to those ways of institutionalization, it is important to notice something else, and that is the dual relationship between certain conceptions of space and of the mind and mental processes. When I say that there is a dual relationship, I mean the following. Within the traditional common sense it has been common to hold, after Descartes, that ideas are "in the mind" in the same way that objects are "in space"; here the relationship of object-to-space becomes a model that seems to support a homologous relation of idea-to-mind. At the same time, within the mind it is assumed that mental operations can be mapped onto the world, so there is a sense in which the relationship runs in the opposite direction. Although the new common sense seems on its face very different and people like Richard Rorty have taken to task epistemologies based on a mapping or mirroring image, there nonetheless remains an important sense in which these images remain powerfully alive.[37] And this is in the notion that one can say "what a work says" and map that onto "what the author believes."

As we shall find in the chapters that follow, it is not all that easy to abandon that view; it is supported not simply by habit and a belief by some that it is true but also by a series of institutional structures. But the point that I wish to make here is more general and ought, from what I have earlier said about the nature of notions of space and their relationship to the ways in which places are constructed, to be clear: this way of thinking about the relationship among what is said, what is believed, and what simply is looks like the product of theoretical engagement but in fact consists of a series of images that are used to tie together disparate phenomena and to simplify that which is otherwise too complex to be managed.

This appeal to these images is, though, a critical error, which directs much useful empirical work in a way that is much less fruitful. But what is the source of the error? Its import has been seen by a number of social theorists, ones as diverse as Peter Winch, Michel de Certeau, Anthony Giddens, and Pierre Bourdieu[38] as well as by students of the sociology of science such as David Bloor, H. M. Collins, and Andrew Pickering.[39] But because they have derived their argument from Wittgenstein, it will be more useful to begin there.[40]

Central to the understanding of these problems is Wittgenstein's analysis of what it means to follow a rule, and he begins with the following example. If someone asks me to add two to a large number and finds that I can in every one of a few tries come up with the correct answer, it seems plausible to imagine that I in fact "know how to add two to a number." But imagine, he suggested, that we teach a pupil the practice of

> writing down a series of the form
> 0, n, 2n, 3n, etc.
> at an order of the form "+n"; so at the order "+1" he writes down the series of natural numbers.—Let us suppose we have done exercises and given him tests up to 1000.

Everything seems to be going swimmingly. But, the example continues:

> Now we get the pupil to continue a series (say +2) beyond 1000—and he writes 1000, 1004, 1008, 1012.
>
> We say to him: "Look what you've done!"—He doesn't understand. We say: "You were meant to add *two*: look how you began the series!"—He answers: "Yes, isn't it right? I thought that was how I was *meant* to do it." (*PI*, I, § 185)

The immediate response—he has not really grasped the rule—seems a plausible explanation here. Yet when we begin to attempt to spell out what this might mean, matters turn out to be by no means as simple as they might have appeared.

The most obvious thing to be said here, the first reaction, is often built on the claim that "When I said do this, I had in mind that at 1000 you should go to 1002." Yet as Wittgenstein argues, surely no one would claim to have had *every* possible item of a series in mind when teaching a rule. A second alternative is to take our minds to be computers and to take the learning of a rule to be very much like programming a computer. We teach the computer a rule by issuing commands or by changing wiring schemes, and then feel confident that if we enter a series of numbers in the appropriate way, the answers will pop out, as though they were foreordained, built in.

But although this seems the commonsense view, it nonetheless presents certain difficulties. For one, people notoriously make mistakes. Further, this view merely sets the question back a step. How, after all, are we to know when we "learn the rule" that the person teaching it means what we think that he or she meant? Perhaps the person teaching us to "add two" will turn out to want us, after 1000, to do what we would call "adding four." Indeed, if we imagine the determination of the application of a rule as a matter of coming up with the correct interpretation, we shall always be left both with the possibility of other interpretations and with the need in interpreting to apply other rules, which need in turn to be interpreted, and on ad infinitum.

One alternative, though, has been to see the development of ways of acting that we normally term "following a rule" as a matter of coming into accord with some external set of social or cultural standards. As Saul Kripke put it,

> There is no objective fact—that we all mean addition by '+', or even that an individual does—that explains our agreement in particular cases. Rather our license to say of each other that we mean addition by '+' is part of a "language game" that sustains itself only because of the brute fact that we generally agree.[41]

But there are two things to note here. First, the appeal to "brute fact" seems to suggest that relativists have been quite right, and that we could, in fact, believe anything that we wished. I will come back to this. And second, and as Michael Lynch has argued, once we accept that the application of a rule rests on something outside of the rule itself, we open the way for causal accounts of the sort widely used in the sociology of scientific knowledge: "Orderly calculation thus depends upon the social conventions we learn through drill; conventions which are inculcated and reinforced by normative practices in the social world around us."[42]

This view does, in some ways, sound quite consistent with Wittgenstein's own way of speaking about the matter. But in fact, and as Lynch points out, it misses his central point, which is that as long as we imagine that there is something called "a rule" and something else called "an application," we shall have the gravest difficulties in attempting to descry the relationship between the two. And this means that "if language is to be a means of communication there must be agreement not only in definitions, but also (queer as this may sound) in judgments" (*PI*, I, § 242). The point is that

> to obey a rule, to make a report, to give an order, to play a game of chess, are *customs* (uses, institutions).
>
> To understand a sentence means to understand a language. To understand a language means to be a master of a technique. (*PI*, I, § 199)

Indeed, "Our language-game is an extension of primitive behaviour. (For our *language-game* is behaviour.) (Instinct.)" (*Z*, § 545). Furthermore, Wittgenstein continues, "you must bear in mind that the language-game is so to say something unpredictable. I mean it is not based on grounds. It is not reasonable (or unreasonable). It is there—like our life" (*Z*, § 559). So "we don't start from certain words, but from certain occasions or activities" (*LCAPRB*, 3):

> Here the term "language-game" is meant to bring into prominence the fact that the speaking of language is part of an activity, or of a form of life. (*PI*, I, § 23)

So turning back to the issue of rules, to the query, But didn't I already intend the whole construction of the sentence (for example) at its beginning? the reply is

> But here we are constructing a misleading picture of "intending," that is, of the use of the word. An intention is embedded in its situation, in human customs and institutions.... In so far as I do intend the construction of the sentence in advance, that is made possible by the fact that I can speak the language in question. (*PI*, I, § 337)

Here Wittgenstein is explicitly rejecting the notion that language can be seen as a sort of lattice, with words combining into simple sentences and gradually combining into more and more general ones. Indeed, philosophical problems arise when we fail to see this and believe that we can apply the evidentiary rules that we use in determining the correctness of "This weighs five pounds" to "The external world exists." Problems arise, that is, from failing to see that language is a complicated, messy human creation, one deeply embedded in human activities.

This view, central to Wittgenstein's later works, developed as a reaction to a very different earlier work, which can be seen in some important respects (but not in all) as a formulation of the traditional common sense. The earlier work was in some ways very much like that of the logical-positivist Vienna Circle, a group of philosophers led by Moritz Schlick and established soon after the turn of the century. Indeed, they saw their work as reflecting the interests and criticisms of two Viennese, Ernst Mach and Wittgenstein himself.[43]

Mach had been the severest sort of critic of metaphysics; he had seen even the assertion that atoms exist as the rankest nonsense and had countered that in order to escape being metaphysical, science must be based strictly on experience. Wittgenstein's early *Tractatus Logico-Philosophicus* appeared to give another kind of criticism of metaphysics. Attempting to understand the relationship between the world and representations of it, he had argued that "the world is the totality of facts, not of things," and "the facts in logical space are the world" (*T,* § 1.1; § 1.13). He had gone on to say that "a picture has logico-pictorial form in common with what it depicts," and "a picture represents a possible situation in logical space" (*T,* § 2.2; § 2.202).

Whereas Mach had offered a critique of the notion that there are objects beyond our perceptions, Wittgenstein appeared in the *Tractatus* to be suggesting a scheme for organizing the data that one did have, a scheme that saw linguistic expressions as having a logical form that mirrored or "pictured" that which existed: "A picture agrees with reality or fails to agree; it is correct or incorrect, true or false" (*T,* § 2.21). These notions had a number of consequences. They established the groundwork for emotivist theories of ethics, where ethical judgments are seen merely as expressions of opinion. And they undercut the possibility of historical works, averring that historians create only what Carl Hempel was later to call "explanation sketches," mere pieces of real explanations, which would necessarily involve laws and look very much like the explanations invoked by physicists.[44]

Although when he wrote the *Tractatus,* Wittgenstein believed that in it he had solved the problem of representation once and for all, by the late 1920s Wittgenstein had begun to have his doubts. The standard story is that he saw the futility of his *Tractarian* view during a confrontation with the economist Piero Sraffa:

> One day . . . when Wittgenstein was insisting that a proposition and that which it describes must have the same "logical form," the same "logical multiplicity," Sraffa made a gesture, familiar to Neapolitans as meaning something like disgust or contempt, of brushing the underneath of his chin with an outward sweep of the fingertips of one hand. And he asked: "What is the logical form of that?" Sraffa's example produced in Wittgenstein the feeling that there was an absurdity in the insistence that a proposition and what it describes must have the same "form." This broke the hold on him of the conception that a proposition must literally be a "picture" of the reality it describes.[45]

It seems unlikely, of course, that this was sufficient to incite such a change. Indeed, K. T. Fann (among others) has argued that what was decisive was a period during which Wittgenstein was an elementary schoolteacher: "The reality of teaching children how to read, write, calculate, etc., and the experience in compiling a dictionary for elementary schools must have contributed to Wittgenstein's later pragmatic view of language."[46]

Central here was first a radically different image of language:

> Our language can be seen as an ancient city: a maze of little streets and squares, of old and new houses, and of houses with additions from different periods; and this surrounded by a multitude of new boroughs with straight regular streets and uniform houses. (*PI*, I, § 18)

Before, language was seen as something either orderly or messy, and orderly language was meaningful language. Assertions that were not in the proper order were beyond meaningful discourse. But now language was seen in a very different way. The new boroughs, science and mathematics, might very well be neat and orderly, but the older parts of language, common expressions and everyday speech, were messy but equally able to inform and communicate.

This image makes the human studies less easy to characterize:

> What determines our judgment, our concepts and reactions, is not what one man is doing now, an individual action, but the whole hurly-burly of human actions, the background against which we see any action.
>
> Seeing life as a weave, this pattern (pretence, say) is not always complete and is varied in a multiplicity of ways. But we, in our conceptual world, keep on seeing the same, recurring with variations. That is how our concepts take it. For concepts are not for use on a single occasion.
>
> And one pattern in the weave is interwoven with many others. (*Z*, § 567–69)

And it is this that renders so difficult the radical change of views. Here views are difficult to maintain because truly to hold them would require that one change not just one, but a great many parts of one's life. Suddenly to adopt the view that all truth is relative to the interests of a particular group, and that one can never "truly know" what members of another group are saying, would be very much like suddenly adopting the view that automobiles are animate objects, or like giving up alcohol after many years of heavy drinking.[47]

If this image of language as a set of practical activities embedded in everyday life, a view that requires that the spatial imagery of the traditional common sense and the new common sense be abandoned, is so compelling, what prevents its more general adoption? Or perhaps more to the point, why is this set of views, developed in order to resolve philosophical difficulties, relevant to everyday life? Some philosophers, and notably Rorty, have argued that in fact philosophy is an enterprise only for philosophers.[48] The suggestion is that the sorts of issues that Wittgenstein raises are not relevant to the practice of science, let alone the everyday concerns of the common person. And there is, it is true, a sense in which the average person is unlikely to develop an abiding concern over issues such as whether other people really have minds, or whether reference is the only legitimate means of fixing meaning.

Further, Wittgenstein himself suggested that the aim of philosophy ought, far from creating architectonic structures to solve problems, to be to dissolve those problems. He commented, "What is your aim in philosophy?—To shew the fly the way out of the fly-bottle" (*PI,* I, § 309). And he said, "The real discovery is the one that makes me capable of stopping doing philosophy when I want to.—The one that gives philosophy peace, so that it is no longer tormented by questions which bring *itself* in question" (*PI,* I, § 133).

Yet in other areas he suggested that philosophical errors are committed not only by philosophers but by everyone. Anthony Kenny quotes him as saying, "Philosophy is a tool which is useful only against philosophers, yes, but also against *the philosopher in us.*"[49] Continuing, Kenny suggests:

> And there are a number of indications that suggest Wittgenstein believed philosophy to be an unavoidable part of the human condition.... In the Big Typescript he says: "Philosophy is embodied not in propositions, but in a language" (MS 213, 425). The philosophy which is embedded in our language is a bad philosophy—it is a mythology. Wittgenstein says, "In our language there is an entire mythology embodied" (MS 213, 434). He gives an instance of what he means: "The primitive forms of our language—noun, adjective, and verb—show the simple picture to which it tries to make everything conform."[50]

Indeed, Kenny continues,

> there is one very important sense in which there could not be progress in philosophy. This is because philosophy is a matter of the will, not of the intellect. Philosophy is something which everybody must do for himself; an activity which is essentially, not just accidentally, a striving against one's own intellectual temptations.[51]

And in fact it makes sense here to see Wittgenstein as having advocated a view of philosophy that shares with feminist critiques the view that an important task is the recognition of the places where we are overwhelmed by images of the accepted and acceptable, and where those images prevent us from seeing what is going on. Indeed, there is good reason to see the process of the development of fresh and new images, the gradual solidification of those images into what are taken to be literal representations, and then the taking of those images as adequate representations of what was once seen in a very different way as a very general process. Certainly, as we saw in the case of Plato, it is one that has long been subject to comment, and one that has long been seen as a source of error and bewitchment. And it is clearly a process that today makes it all the more difficult to understand the nature of the written work.

Toward a Geography of the Work in the World

What I have suggested here is that a certain style of thinking prevents our seeing what is happening in the world, and in the matter at hand prevents our seeing the workings of the written work. We might want to characterize this style as one that imagines thinking to be a matter of the mental application of abstract rules, where the mind is seen as being very much like a computer. We might want to characterize it as a style in which the particulars that make up the world are imagined as existing within a large, abstract container called "space." And we might want to characterize it as a style in which the mind and thought are seen on the image of space and the particular, with thoughts floating around in a mental space in the very same way that planets and asteroids float around in physical space. This style of thinking is very much in evidence in the codifications of the traditional common sense, and it ought to be clear from my characterization of the new common sense that its proponents—postmodernists, anthropologists, rhetoricians—have seen their task to be the effacing of that view and its replacement with a better one.

Had they succeeded, we might very well be satisfied with the answers provided in the new common sense to the questions raised about language, knowledge, existence—and the written work. And yet, I have suggested that to the most obvious of questions, like, Why is it so hard to change one's mind? and, If all knowledge is relative, why don't people act as if it is? the new common sense provides thin gruel. And this is because it has remained firmly in the camp of the traditional common sense, in the way that it has remained loyal to these defining styles of thinking, about space and place, about the mind and its thoughts, and about the application of theory and rules.

In a fundamental way proponents of the new common sense have found it difficult to rethink these issues. And so, the view that a geographer has a philosophical or theoretical "position" in conceptual "space" remains; the view that an author can be seen as "having a position" that can be summarized, abstracted, characterized also remains. The view that a book can be characterized as "saying something" remains. The view that a book is a mapping of the author's ideas remains. The view that one's conclusions follow inexorably from one's premises, and in a way that can be made public for all to see, remains. And perhaps most central, they retain the view that one can simply change one's mind about this or that, and then go about one's business.

That the codifiers of the new common sense have held these views is no surprise. Each of these views or positions remains deeply held, and for reasons that Wittgenstein laid out. For one, there is the difficulty in coming up with a more adequate way of talking; after all *I myself* have just characterized the view that people have views and positions as itself a view, or a position. But more to the point, to "hold a view" is not just to have a set of ideas in one's mind. Rather, to hold a view or belief is to engage in a range of actions. It is to say and do certain things at certain times and at certain places. To change one's views or beliefs is not merely a matter of turning in the old set for a new model. Rather, it requires that one change one's actions, and in doing so, change the ways that one engages other people, and the places in which one engages them. Indeed, it is to change the place in which one lives and acts.

And such changes, even when we wish them, need always to negotiate the institutional and technological landscape in which we live and act. But that landscape is itself rife with images that tell us what can be and what cannot, what can be believed and what cannot. We live, in part, in a landscape that constantly tells us that the universe is a Newtonian one, and that equally tells us that the Aristotelian vocabulary of places and belonging and fitting lacks the institutional imprimatur that it needs.

We have seen that the written work, even while seen as primarily a repository of ideas, has also been seen as an element of a system of authority, as a commodity, and as an element in a classificatory system. By being involved in each of those systems, the work is imbricated in institutional and technological settings that tell those who make and use it what they are doing. There is simply no denying this fact. And it is this to which we must turn if we are to gain a better appreciation of "what a work says," because what any work says to a reader derives not simply from the word on the page but also from the fact that the word is *on* the page, in a book, edited, published, cataloged, reviewed, cited, and criticized. All of these activities take place in time—and in place.

PART III

The Work in the World

CHAPTER FIVE

Authorship and the Construction of Authority

In a well-known article, Michel Foucault described what he termed the "author function."[1] "The coming into being of the notion of 'author,'" he said, "constitutes the privileged moment of *individualization* in the history of ideas, knowledge, literature, philosophy, and the sciences" (141). According to Foucault,

> The text always contains a certain number of signs referring to the author. . . . Such elements do not play the same role in discourses provided with the author-function as in those lacking it. In the latter, such "shifters" refer to the real speaker and to the spatio-temporal coordinates of the discourse. . . . In the former, however, their role is more complex and variable. In fact, . . . all discourses endowed with the author-function do possess this plurality of self. (152)

In literature the author function developed only recently. In contrast, if we turn to works of science, we find that

> those texts that we would now call scientific — those dealing with cosmology and the heavens, medicine and illnesses, natural sciences and geography — were accepted in the Middle Ages, and accepted as "true," only when marked with the name of their author. . . .
>
> A reversal occurred in the seventeenth or eighteenth century. Scientific discourses began to be received for themselves, in the

> anonymity of an established or always demonstrable truth; their membership in a systematic ensemble, and not the reference to the individual who produced them, stood as their guarantee. The author function faded away. . . . (149)

For Foucault the development of this author function results from several processes. For one, we only have an author function after it becomes possible for works to be "transgressive," and for an individual to be taken to be responsible for the contents of the work. At the same time, the author function is "the result of a complex operation which constructs a certain rational being we call 'author' " (150). He concludes that what is needed is an answer to questions such as "How, under what conditions and in what form can something like a subject appear in the order of discourse? What place can it occupy in each type of discourse, what functions can it assume, and by obeying what rules?" (158).

We saw in the formalization of the traditional common sense elements of both of these processes. We saw in Descartes what Foucault has viewed as a move away from the appeal to the author function in science, as the validity of a work came to be seen as arising not from its author's status but rather from something about the work itself. We saw at the same time that contemporary notions of intellectual property developed in a historical moment in which printers wished to claim rights to works, and in which the view of the right to a work as deriving from the labor invested in it allowed those printers to claim rights to works that in the past would have been defined in terms of that author function.

But the conception of the "author" is in modern-day geography more complex than the reference to this pair of developments might indicate. In fact, in order to understand the place of the subject in discourse, we need to see that we actually view any text as having multiple functions and its author as filling multiple roles, where these roles are at times supported and made visible and at others rendered invisible by images and technologies of support and effacement.

In this chapter I shall consider three very common views of the author and one that is far less obvious. In the discussion of the first three I shall point to the ways in which a set of images and technologies has lent them visible support. I shall point in the discussion of the fourth to the ways in which its role has been *obscured* by a set of images and technologies. Finally, I shall turn to the work of a particular geographer and note the ways in which we find each of these conceptions of authorship.

The Everyday Author

TEXT AS REFLECTION

Today a highly respected form of text is that which claims to be a reflection of the world and especially of processes that underlie the world. In one sense virtually any explanation claims to be such a reflection, but some make this claim in more obvious ways. As we saw, from a modernist point of view this image of the text has its recent roots in Descartes, in the view of the text as a reflection of the ideas held by the author, and then of the world. On this view a text may be seen as linguistic or not; models, whether mathematical or not, photographs, block diagrams, and certain kinds of maps all attempt to reflect the world in this way.

But even in cases where models and the like are not at the fore, the structure of a text may itself lead the reader to the view that what is being propounded is a reflection of the world. Various stylistic means are used to lead the reader to this view. For example, certain styles of citation suggest that knowledge consists of larger elements and structures that are built upon more basic ones; these styles—the Harvard style is the most obvious—suggest that one can create a mapping from the structure of the text to the structure of the world. Further, an article that moves from abstract to introduction, hypotheses, methods, and conclusions suggests that it is an element of a larger structure.

Thus in one sense to see a text as reflection is to see it as a reflection *of* the world. In another sense the text may be seen as a grid or template *through which* one observes the world; there the text suggests that through it one may see a world stripped of contingent elements. A mathematical model, for example, characterizes the world in terms of an abstracted and restricted set of elements and relationships—along with the admonition that these are what is important. A block diagram does much the same. Although more difficult, in a sense photographs are quite similar. They require more training in order for a student to be able to read them in a way that omits the extraneous, but because they seem to have almost a mechanical relationship with the world, they need certainly to be seen as a kind of lens on the world.

Moreover, to say that a text is a reflection is not necessarily to see it as an *accurate* reflection. It is, after all, with the new common sense possible to imagine that the written work is a direct reflection of the interests of the author rather than of its explicit subject matter.

TEXT AS CATALOG

The text as reflection is the function of scientific and geographic works that is the best known and that is seen as least problematic. And as we have seen, this view is canonized in discussions of the nature of science. But it is not the only one, or even the only common one. It is also common to imagine the writing of a work as a matter of the creation of a catalog. By this I mean, simply, that some geographic texts are (or contain) catalogs of facts or information. There are a wide range of examples within the discipline; indeed, many outside of geography would see what we do as primarily a matter of creating catalogs, of rivers and lakes, principal products, and so on. Some maps—I have in mind here especially historical atlases—are explicitly intended to be catalogs of information. We find catalogs, too, in charts and tables within texts.

I would note two things about this form of text. First, it brings with it a kind of authority that the others will turn out to lack. However much one may hear arguments that factual statements are "theory laden," there is a kind of authority to the statement "Coal production dropped 25 percent last year in China" that is denied to other kinds of statements. When those statements refer to objects in a kind of middle range, ones neither micro- nor macroscopic, the fact that one might in principle go out and look at the objects and events carries a weight that can be denied only by those who do not go out and look.

Second, with respect both to their audiences and to the social world more generally, catalogs have a kind of value that is in some respects quite different from that had by other kinds of texts. In many instances one's response on finding that information that one needs has already been collected is, "Thank God, it saved me the work of doing it myself." Catalogs, that is, are valued because the person who created them has invested labor in them.

But note that here, too, one need not believe that the facts cataloged in a work somehow exist naively "out there" in the world in order to see a work as a catalog; one may have the very same view, but grounded within the new-common-sense notion that facts are constructed by an author and reflective of that author's interests.

Within geography there has been a tendency to denigrate these works or portions of works, a tendency to see them as not very important. They have taken on for geography the role that chronicles have taken on for history, the role of primitive precursor. This denigration has happened in one way in the quantitative revolution, and then in another in the later postmodern turn. Nonetheless, one would be hard pressed to find a work that is not in some respects such a catalog. And in fact, there are some areas, like history, development studies, and geographic information systems, where the acquisition of data remains an important end in itself. This is the case in history, where historians who only interpret but do not have their own sets of original data are looked down upon. It is the case in development studies, where a failure to have gone into the field to collect data is similarly seen as evidence of a lack of seriousness. And at the center of geographic information systems are collections of just such data.

TEXT AS ACHIEVEMENT

There is a third sort of text in geography, one that has some similarity to the text when conceived as a reflection created by the scientist or geographer. Here, though, the central issue is the relationship not between the text and the world but rather between the geographer and the text. Here the text is seen fundamentally not as a reflection of the world but rather as an achievement of the geographer. So rather than being asked to look at the world through the text, we are asked instead to look through the eyes of the geographer, a person who has developed a particularly powerful way of seeing.

There are a number of ways in which this sort of view appears, and a number of ways of identifying it. We see it where the author presents the work as part of an ongoing personal travail, as a step on the road to enlightenment. This presentation may involve directly personal statements about the author's difficulties and triumphs.

We find this view, too, as we consider the rhetorical structure of written works. In an important sense, of course, we see this where the author presents a work as the result of a struggle along the path from ignorance to knowledge. It is important to see that this view does not occur merely in the sort of autobiographical works that we associate with certain humanist geographers. Indeed, it is often codified into style manuals, where the author bravely presents a hypothesis, searches out data, and ultimately is vindicated as the data support what was once seen as an opportunity as much for failure as for success.

Language is often at the center of attention of such texts; the author may claim that language is simply inadequate, that we need better or more powerful means of linguistic expression. Almost defensively, neologisms are invented; parentheses begin to appear in the midst of words. The rules of punctuation are violated, even flouted. On the other hand, the evidence of this sort of text may lie in the success of language, in the memorable metaphor.

Now whether we are considering the text as a kind of bildungsroman in terms of the author's personal struggle, as a rhetorically structured work demonstrating the move from ignorance and chance to knowledge and certainty, or as a vehicle wherein the limits of language are tested and felt, there is an important sense in which we are seeing the text as having not merely a contingent but an essential and inextricable relationship to the author and, indeed, to the author's very existence. And because of this view, this version of authorship, so common elsewhere, is seldom acknowledged in the geographical work, except in that limited realm of the confessional, in works like Peter Gould's *The Geographer at Work*, Peter Haggett's *The Geographer's Art*, and Allan Pred's "The Academic Past through a Time-Geographic Looking Glass."[2]

The Technologies of Support

These common views of the nature of authorship are supported in a number of ways. As we have already seen, to the extent that a work is seen as someone's product, one may be justified in declaring that person the author. I shall return to this issue later in this chapter because the pervasiveness of theories of property is a key feature of the effacement of one view of the author. But here I should like to turn to two ways in which these views of authorship are maintained, first through the formalization of systems of style, and second through the formalization of systems of citation. Each of these ways strongly supports a view of the author as the creator of a reflection, a catalog, or both. And each, too, is right at the heart of the difficulty that authors have in attempting to express new and different views of the world.

THE FORMALIZATION OF STYLE

We have seen that the publication system of geography and science is increasingly made up of units, journals and articles, classified works, and commodities. Moreover, the written works themselves, the articles, books, and reviews, have themselves come increasingly to be internally differentiated. One way has been through the formalization of requirements for writing, and especially through the adoption of style manuals. The use of style manuals has become all but ubiquitous, and those style manuals have been univocal, if unselfconscious, in their expressions of particular norms for publishing behavior. The notion of a style manual is actually relatively recent. The earliest style manuals were *Hart's Rules for Compositors and Readers,* published in 1893, and *The Chicago Manual of Style,* published in 1906. The first manuals directed at scientists were Clifford Allbutt's *Notes on the Composition of Scientific Papers,* published in Britain in 1904, and Trelease's *Preparation of Scientific and Technical Papers,* published in the United States in 1925.[3] As recent as they are, such manuals are now ubiquitous; in geography, as elsewhere, journals publish their own manuals of style, often changing them upon changes in editor or publisher.

In part, the style manual is a way in which the journal defines the relationship between the author and the work. The *Annals of the Association of American Geographers,* for example, recommends that an author use the active voice. This requirement appears to mandate documents that through the use of that voice express the view that the author is "on top of things," in control, as it suggests the constant interjection of the persona of the author. At the same time, though, this requirement "flattens" the text, to the extent that it makes the active voice the *only* voice and thereby removes any sense of agency from the discourse involved. The requirement has a second effect: in the ongoing debate about geography as "art or science," it makes a choice for the author.[4]

More fundamentally, whatever the requirements that these manuals specify, they express the journal's vision of ideal communication, and in doing so insist that that vision be shared.[5] In that sense, the standardization of style needs to be seen as part of the establishment of a form of discipline within the discipline.[6] That this is the case is made forcefully obvious by the shrillness of the responses to ex-editor Asa Kashar's "Style! Why Bother?"[7] His commentators were almost unanimous in deploring his view that there is no reason for journals to insist on uniform, rather than merely comprehensible, citation and reference systems.[8] They were as unanimous in being unable to say why they found his suggestions so disturbing.

But style manuals have a second function, and that is to organize and to define what can count as intellectual property. Here again the *Annals* has been typical of trends occurring throughout the discipline and beyond, and an analysis of changes in its style manuals since 1964 suggests that this process of definition has occurred in a number of ways. First has been through the adoption of the requirement that each article be accompanied by an "abstract." This requirement, originally instituted in 1964, appears benign. Yet it has the effect of privileging certain sorts of works, those that appear capable of being abstracted in a mechanical and unambiguous way through characterization of data, methods, and findings. Here, in part, the requirement that there be an abstract suggests that the fundamental function of an article is to provide or disseminate information in the form of facts and findings or theories and hypotheses rather than to advance interpretations or narratives.

The requirement equally suggests that the abstract in presenting the "essence" of an article thereby defines it as a piece of intellectual property; items by the same author that can be abstracted identically are by definition identical. Requiring an abstract, of course, has a series of practical consequences; among them, it plays into the hands of those who wish to find out what an article contains without having to read it, and among the most important of these people are not practitioners of specific disciplines but those involved in the cataloging and collating of information. (Much the same could be said of the later requirement that articles be accompanied by a series of "keywords" to be used for indexing; ironically, the director of the Institute for Scientific Information, which produces citation indexes, has argued that keywords provide a decidedly *unscientific* basis on which to index articles.)[9]

A second major style change occurred in 1982, when the *Annals* moved from a referencing system that used footnotes, one based on Modern Language Association standards, to the so-called Harvard system, wherein references are made parenthetically within the text.[10] At one time an author was able to use footnotes to make editorial asides, stepping out of his or her role as a "neutral" reporter to address the reader more directly or informally, and to engage in a kind of dialogue, but now such notes are discouraged. Neutral as this change may seem, it has several effects. Perhaps most obviously, it virtually precludes the publication of works that rely heavily on archival sources; in that way it skews publication away from historical subjects and benefits contemporary works and works of secondary scholarship. The change also makes it easier for readers to "mine" works, because they no longer need go through references one by one. But more important, it neutralizes citations, making each of equal weight, and thus makes it much more difficult for an author to qualify his or her conclusions or his or her references. Here the *Annals* and other like-minded journals have adopted a new notion of authorship. As Robert Day put the matter in *How to Write and Publish a Scientific Paper,* "A scientific paper is not 'literature.' The preparer of a scientific paper is not really an 'author' in the literary sense. In fact, I go so far as to say that, if the ingredients are properly organized, the paper will virtually write itself."[11]

This notion of authorship, where a paper consists of an admixture of "ingredients," thereby defines intellectual property in terms of those very ingredients. As the prohibiting of footnotes and asides "flattens" the text, the purpose of an article becomes simply to "report findings," to provide information. At the same time, the use of the Harvard system makes it easier to implement a systematized program of rewards.

These changes in the journal occur against a background in which authors appear to be increasingly self-conscious. We find evidence of this in a number of areas, one of which is the development of "how-to" manuals for young practitioners. In geography the most recent of these how-to manuals consists of a series of articles published in an edited volume.[12] There professional geographers have offered their younger and aspiring colleagues advice on making their ways in the field and on being published and recognized. For example, on the basis of Stanley Brunn's characterization of geographers as falling into one of five categories, from highly productive to professionally inactive, David Butler avers that "you cannot afford to be perceived as any lower than group-two status."[13] Although he believes that publishing leads to "the knowledge that you have contributed to your discipline and perhaps to scientific advance in general," he prefaces that statement with the assertion that "you simply must publish if you wish to be promoted and tenured in a university setting."[14] In fact, his and other recent articles make no bones about being concerned in the first instance with teaching scholars not how to do better work but rather how to do work that will be accepted.

Butler and others in the same volume make a number of other suggestions. One needs to write, actually, to a very small audience—the editor and the group of possible referees. Works ought to be structured in the way that one's audience will expect; in physical geography this may require a description of the study site, but in human geography it may involve historical background or even photographs. In any case, the various elements of a written work ought to be organized in a standard way, which in all but some historical and semipopular journals will include an introduction, description of the problem, description of methods, results, and conclusion.[15] Further, in writing to an audience one ought to adopt a style that will be seen as neutral, while recognizing that what is counted as neutral will vary from journal to journal. Finally, one ought to include appropriate references. In part, of course, these will be determined by the intellectual debts accrued in the work, but it is also important to include references that the presumed audience—editors and referees—will expect to see.

These how-to manuals express a newly public sense of self-consciousness about the discipline. But from the point of view of one interested in judging the work of others, this has unfortunate consequences. To the extent that we believe that authors have taken seriously these strictures, we no longer feel comfortable "reading off" the traditional set of Mertonian norms from their behavior. We feel as though we may more easily than in the past be deceived, that we are looking not at a piece of science but at an advertisement for the author.

Hence we see in the journal today an attempt by editors and publishers to maintain a sense of order and discipline, paralleled by an attempt by others to second guess them, to manipulate the system for their own advantage by appealing directly to the needs and desires of those editors and publishers. Here the standard system that Merton described seems truly to be coming apart. The motivations of each of the actors appears now to be quite foreign to those which he described, just as the institutional structures that supported disinterestedness appear no longer to be functioning as efficiently as they once did. Under allegations of bias, journal editors go on the attack, just as do publishers told that their wares are too expensive. How-to manuals suggest ways in which young practitioners can look in the first instance to success rather than truth; the pretense that one simply does the best possible work and lets matters of status sort themselves out is gone. The establishment of style manuals has only amplified this tendency; at the same time, it has codified a notion of intellectual property, where the basic task of an author is to deal with facts and findings, and where ambiguity, hesitancy, and humility have no place.

THE TECHNOLOGY OF CITATION

One central element of the style manual, and a central one here, has been the codification of the ways in which authors give credit. In a sense this technology is even more pervasive in its support of the views of the author as cataloger and reflector than is the style manual itself, and as we shall see, this is because the process of citation has become not merely more formalized but also more strongly articulated within the larger system of publishing and the measurement of scholarly authority. In the past the process of citing was informal—to a point that in French scholarship, for example, one found difficult to fathom—but it no longer is. I mentioned earlier that in the past several years journals like the *Annals* have changed their citation practicing, replacing a system based on footnotes with one based on references embedded in the text. There are no doubt a number of reasons for this change; some would mention economics, others that the new system "looks more scientific."

But whatever the reason, the new system has had an unexpected consequence; it has eased the introduction of formalized means for the measurement of "quality." It has allowed the measurement of the quality of an article or book in ways that support the image of science that we have already seen at work in other ways. Indeed, it seems in a very real sense to have created a system within which written works need no longer be read for and judged in terms of their content. Rather—and certainly for those whose primary interests have never been in the content of works—it now seems possible to move away from that subjective approach and adopt one that is more "objective." In doing so, however, it is possible to separate utterly the levels of rewards and communications. This has happened through the development of citation analysis.

The development of first hard-copy and then on-line and optical-disk versions of indexes for the sciences, social sciences, and humanities has provided a new and quantitative way of dealing with traditional questions of productivity and quality, and hence of measuring the functioning of the reward system within science. Yet this interest has not been directly accompanied by an understanding that it involves a recasting of the system within which the journal operates.

Citation indexes have spawned a huge literature, both outside of geography and within it.[16] In geography this work has been relatively straightforward and has for the most part simply involved the counting of citations and the ranking of departments and individuals,[17] although there have been forays into more complex forms of analysis.[18] All analyses, however, rest upon the view that footnotes and citations fulfill a dual role. On the one hand, they act as the intellectual foundations for a work; they point the reader to sources for data or concepts and thereby relieve the author of the need to repeat what has already been done.[19] On the other hand, they become a means for recognition of the ownership of ideas; in principle, an author in his or her notes gives credit for all of the intellectual property embodied in the work.[20] In the latter sense footnotes and citations are embedded in an economic model of science, where ideas are in the first instance property to be owned.[21]

Although journals have been around for more than three hundred years, it is not until about 1850 that we first find a modern version of the use of citations and references.[22] And although there had previously been a long tradition of footnoting, and although since Bacon the view of intellectual development as a matter of the gradual accumulation of facts had been common, it was only then that the citation gained its present role as a means of making claims to intellectual ownership rights.

Even so, there were barriers to the development of the use of citations either as a clearly articulated foundation for science or as an equally articulated statement of the ownership of ideas. The development of a systematic approach to citation and referencing met stylistic barriers; before the publication of standardized style manuals, formats for citations, as for other aspects of publishing, were decidedly personal in flavor. Another barrier may have been custom; well through the nineteenth century the primary outlet for publication remained the book, and books comprise a notoriously complex system of communication. But the more important reason for the relative recency of the development of this system may simply have been that there were not very many scientists, and that specialization was achieved only relatively late. According to Derek de Solla Price, in the United States there were only about 1,000 scientists in 1800, 10,000 in 1850, and 100,000 in 1900.[23] And in 1800 there were only about one hundred substantive journals in the West.[24] It was not until early in the nineteenth century that scientists began to professionalize and specialize, and hence it was only then that the issue of ownership came to be conceptualized as it now is. Even then, in the context of what Price termed "little science," the formalization of that ownership system was not really necessary.

The period between 1850 and 1915 was in some ways the golden age of professional, autonomous science—wherein scientists retained the individuality that went with a lack of major institutional support while they at the same time became assimilated into the university. It was also the period that saw the development of new formal styles alongside a science that discipline by discipline was still relatively small and therefore able to operate through traditional and customary forms of authority.[25] But by the late 1950s and early 1960s, we had entered an era of "big science," which I term the "contemporary," where much of what was begun in the nineteenth and early twentieth centuries had begun to mature. In the United States there were perhaps one million people with scientific and technical degrees; in the world there were 30,000 scientific periodicals, which had produced about six million papers and were producing another half million per year. At this point the changes in the system by which discipline and control are established and maintained appeared to be leading science in new directions.

The very factors that led to the growth of science at the same time led to its becoming increasingly an anonymous group of practitioners rather than a more traditional community. In part this was a response to the sheer increase in numbers, but it was also a response to growing specialization and the increasing practice of science in other than what had only recently become the traditional venue, the university. In this context, and especially because of the increasing demands of extramural funding agencies and of those claiming various forms of discrimination, there were increasing pressures to develop better means of judging the quality of work; it appeared that personal contact and reputation could no longer be relied upon. One way of judging quality or output was simply through the counting of publications, and this had the distinct advantage of being consistent with the contemporary ethos of science. But this approach to judging the quality of work had clear disadvantages. Perhaps most important, the number of outlets for presentation of scientific work had grown far more rapidly than had the number of scientists. The result was that in the natural sciences it became and has remained relatively easy for materials to be published; in an average physics journal, for example, about eighty percent of submitted articles are accepted, and if one considers articles rejected by one journal and accepted by another, the rate is higher. Hence, the mere fact of having been published seems to mean less and less.

In the early 1960s, Eugene Garfield stepped into the breach with his brainchild, the *Science Citation Index*. According to Garfield, the index had a single fundamental advantage over other attempts at managing information in science: through the use of citations, it indexed that information according to categories established by the author, who in mentioning other works and authors directly expressed the location of his or her work within the broader context of science.[26] Hence, the index was established, according to Garfield, using the methodology of science itself, and not according to some transient system of fiats or some "*a priori*" scheme. It thereby escaped the faults of systems based on abstracts, keywords, or titles.

Garfield explicitly claimed that the *Science Citation Index* and the others to follow were designed as means by which to deal with the "information explosion" and thereby, in a sense, to fulfill Bacon's dream of science. Indeed, he argued that it seemed possible that we would soon see the establishment of a world information bank and might "achieve 'total communication,' a state of research nirvana."[27] In the process of moving toward research nirvana, Garfield developed the Institute for Scientific Information (ISI), today a privately held corporation with seven hundred employees and total annual sales of $44 million; it is the sixty-fifth largest publishing firm in the United States, well behind Time-Life's $4.2 billion in sales but ahead of the AAAS, the publisher of *Science* and hence the harbinger of that very nirvana.[28]

Although the primary claim was that citation indexes could be used for research in science, Garfield was quick to point out that they had other possible applications. He asserted that a citation index might be used for

> the evaluation of the impact of a paper, a man's works, a journal, material published during specific time intervals, the works of specific teachers, works coming out of a university or department, work financially sponsored by a specific agency. . . .
>
> It may be used to study journal utilizations, measuring literature habits of scientists, effectiveness of specific journals in reaching specific audiences, purchasing requirements of specific libraries.[29]

And so the journal has become a formal means of surveillance not simply of the author but of a whole range of institutions. And as such, it has tangled within itself very specific notions of the nature of authorship.

The Other Side of Authorship: Text as Dialogue

I noted earlier that there are three main, and well-accepted, images of the written work. But in fact we find a fourth way in which it is common to view the written work, and that is as a piece of dialogue. This may seem an odd thing to say. After all, it is rare indeed to see a work in geography written in the form of a dialogue, and those that are, are often seen as offbeat, strange. It seems on the face of it more reasonable in searching for dialogues in science to turn as far back as Galileo, for whom the dialogical form functioned at least in part as a means for pressing unpopular and even dangerous ideas.[30] Indeed, one might make a case for looking back farther, to the move from Plato to Aristotle, and to what is conventionally seen as *the* move to science from something better called critique. Yet to look at the written work and the process of authoring in this way is to be led astray by the power of the images propounded by Descartes and his followers. For in fact dialogue forms a central part of the written work in science and geography.

The most explicit place where dialogue occurs is of course in commentaries and replies, and these are certainly common in the journal literature. But we also see dialogue in citations. Where in one sense we may wish to imagine the process of citation as a matter of the use of other works as foundations for one's own, in a great many cases one cites another as a means of addressing that person's point of view. Indeed, and especially in footnotes, one can engage in a kind of dialogue with other members of one's field, questioning them here, agreeing with them there. Moreover, in footnotes one can engage in a dialogue with oneself, questioning whether this or that view makes sense, whether this position is overstated, whether one research direction rather than another will be more profitable.

It needs to be noted, too, that one engages in a kind of dialogue through silence. Just as one engages a person or work through citation, where one fails to cite a possible and obvious work, one also engages in a kind of dialogue, a dialogue through silent dismissal. And this is perhaps easiest to see if we think of the process of citing *as* a process of dialogue, in a place; that is, we need to think of citation less as a matter of the construction of a hierarchical structure than as one of an author's engaging with other authors and representing the written work as an outgrowth of that dialogue. Not to be included in that dialogue is to be outside the place wherein it occurs.

The Technologies of Effacement

If dialogue is so important, why is it so much less visible in conceptualizations of the nature of the author? It seems to me that there are really two connected reasons. The first reason for this effacement is that for an author to engage in dialogue, more than to reflect, to catalog, or to achieve, is to be associated with a place; hence, the dialogical function runs counter to that powerful set of images that have elided the difference between space and place. And second, dialogue runs counter to the institutionally entrenched views of the author as an individual and of authoring as a matter of the production of property.

THE MODERNIST IMAGE

The first reason for the difficulty in seeing the dialogical function of the written work derives from the power of the modernist image at work in the traditional common sense and, as we have seen, the new common sense as well. This image is not at work merely in the common appeal to certain theories of representation, although it is that. It is also and more importantly involved in the way in which the appeal to those images denies the importance of the very places necessary to the existence of dialogue. If dialogue is fundamentally associated with places, a theory that denies the importance of places, confounding place and space, at once denies the possibility of dialogue.

If this difficulty were merely a matter of people being somehow captivated by a set of images that obscured the import of dialogue, the matter might be simple; one might need simply to point to what had been overlooked. But in fact these images are themselves supported in a series of ways, and it is the strength of this support that makes it so difficult to change, and or at least see past, these images.

Right at the center of these images is the book or article itself, as a printed and physical object. As such it itself constitutes a kind of fulcrum around which a set of images are organized. First, as a physical object, printed and perhaps bound, it is permanent and univocal; if we all know that a written work is subject to endless interpretation, we also feel that somehow and in some sense that interpretation does nothing to the work itself. By contrast, dialogue is quite the opposite; it is ephemeral, and not permanent. In fact, this is true even of the representations of dialogue that we find in Galileo and the like; these dialogues really are permanent, frozen in time. And this seems to strengthen the view that there is no place for dialogue within a written work.

As permanent objects, written works provide a second image that undercuts the view that they may be dialogical in form; they do so to the extent that they appear to be stating a view. Where dialogue is a matter of constant negotiation between individuals, the written work appears univocal, it appears to contain and then to state facts and ideas. There appears to be no place in the act of reading to inquire of the author, to check on a fact, resolve an ambiguity, and certainly to effect a change of mind.

Finally, and again as permanent objects, the written work is portable—just as dialogue is surely not. One can imagine—although as will become clear in the next chapter, *only* imagine—reading a book anywhere, anytime.

By contrast, dialogue seems fundamentally to occur in time and to be bound to a place.

THEORIES OF PROPERTY

So in each of these three ways the "fact of the written work" appears to be testament to existence of the work as an object of which dialogue has no part. This, though, is related to a second source of the effacement of dialogue; the way in which the existence of the individual, permanent work has led to technological and institutional underpinnings of the views of the text as reflection, catalog, and achievement in the form of regulations related to intellectual property. Together these notions of the author and the written work are supported by a complex of institutions that provide visible structuring of the interactions among the author, the work, and various publics. In an important sense these notions of intellectual property emplace the written work, making it a part of the world.

In Western thought there are two typical and long-standing ways of conceiving of the basis for rights to property. One focuses on what Roland Barthes termed the "author," on the products of creation and insight. Within this tradition the written work is viewed as functioning as a means of defining the personality of the author through the establishment of a persona. And this process, where a person uses objects as a means of defining her personality, is at the heart of an alternative view of the nature of rights to property, a view propounded by Hegel.[31] According to Hegel,

> A person must translate his freedom into an external sphere in order to exist as Idea. . . . (Para. 41)
>
> The rationale of property is to be found not in the satisfaction of needs but in the supersession of the pure subjectivity of personality. In his property a person exists for the first time as reason. (Addition, Para. 41)
>
> Mental attitudes, erudition, artistic skill, even things ecclesiastical (like sermons, masses, prayers, consecration of votive objects), inventions, and so forth become subjects of a contract, brought on to a parity, through being bought and sold, with things recognized as things. (Para. 43)

People, Hegel argues, are not merely separate and separable individuals, and it makes no sense to think of a person as having her own intrinsic personality. Rather, people exist as people only insofar as their ideas and desires are expressed in a public forum. Property is a fundamental mode of that existence, and, indeed, without property a person is simply not a person.

That this view of property is not simply the ranting of a compulsive systematizer can be seen in the way in which this notion has been embodied in legal practice.[32] This is especially the case in the French notion of moral right, which is the more appropriate here because it refers not to property in general but rather to intellectual property, to the works created by artists and writers.

There is an alternative to this personality theory of property, one which has been far more important in American (and English) law than its Hegelian, Mand Continental, counterpart. This view comes from John Locke. In the *Second Treatise of Government* he argued that

> man, by being a master of himself and proprietor of his own person and the actions or labour of it, had still in himself the great foundation of property; . . .
>
> Thus labour, in the beginning, gave a right of property wherever anyone was pleased to employ it upon what was common. . . .[33]

For a Lockean, then, when an individual invests labor in something, that thing becomes her property. This way of looking at the matter is actually consistent with common sense; further, it is one that has long since been incorporated into the law. For example, when a builder works on a building, that person's labor comes to be incorporated into the building. If the owner fails to pay the builder, the builder can put a mechanic's lien on it, and becomes the owner just by virtue of the labor incorporated into it.

Now, the Lockean labor theory of property appears to clarify certain features of two primary means of registering ownership rights, the copyright and the patent. Each of those was established as a means of protecting the works of inventors and creators. But both were established, in the first instance, as means of promoting invention and creation by giving the creator the right to the monetary fruits of her work. That this is the case is demonstrated by two features common to both systems; in both cases it is possible to sell or otherwise alienate one's rights, and in both the rights are of limited duration. Indeed, the role of labor in patentable projects has recently been made clearer in discussions of the duration of patents; some have argued that a patent ought to last only for the length of time that would be required for someone else to duplicate the project.

Here, though, it should be clear that there are features of copyright and patent law that are dramatically opposed to the usual practice within academic science. Foremost, certainly, is the issue of alienation. Under the system of moral right, it would make no sense to alienate the rights to a work; indeed, those rights are passed down through one's heirs, in whom one's name is kept alive. I might buy the copyright to a person's latest opus, but what I decidedly could not do is then say, "OK, we'll now call it Jones (1991) instead of Stoddart (1991)." If we could, of course, tenure and promotion deliberations would be much more interesting; I could change my area of specialization just by writing a check.

Indeed, in French law moral rights are so closely tied to personality that if I later decide that one of my creations no longer adequately expresses that personality, I can demand the return of all copies of it. Further, I cannot be sued for damages if I fail to produce a promised product; to require me to produce it would be to make me express myself in ways that might not be truly me.

In the case of moral right, rights do not expire, and neither do responsibilities. And here, too, the notion is consistent with certain features of commonsense ways of thinking about intellectual creations. Plagiarism, for example, is not excused simply because the copyright has expired on the work being plagiarized. And although we might believe—or even hope—that legal responsibilities for errors or omissions ought to expire, we are less likely to think that the author's responsibility for them similarly expires.

So if it turns out that the labor theory of property has much to say about commonsense views of the justification for property, it also appears that the personality theory clarifies certain features of property rights that are commonly accepted within academic practice. And until recently, in American practice these two views have coexisted more or less peacefully. In academics, for example, the practice of citation has appealed primarily to the personality theory, while at the same time, an author typically gives away the copyright to an article, which is then held by a publisher who appeals to the labor theory for support.

But it needs to be noted here that notwithstanding their differences, these two theories provide underpinnings for the three central notions of authoring as the creation of the written work as a catalog, a reflection, and an achievement, and in doing so indirectly exclude the dialogical elements of the written work, and of authorship. Putting the matter perhaps a bit too starkly, a catalog appears in the first instance to be a product of labor, even a product of labor pure and simple. To the extent that a written work is an achievement, it is an expression of the personality of the author. And depending on one's point of view, to say that a text is a reflection might be to say one or both. To the extent that it is a reflection of the author's mind, it is in a sense an expression of that mind, and hence tied to the author's personality. To the extent that it is a simple reflection of the world itself, though, one might prefer to see it as an object of labor (or even as an object that ought not to be seen *as* a piece of property). But to see a written work as a dialogical object, firmly ensconced in a set of social relationships, is to undercut the entire idea of the object as a piece of property.

On the Practice of Authoring: David Harvey and the Refractory Text

And so when we consider the nature of authorship, we are likely to see a written work as the product of an act of authorship aimed at cataloging, at reflecting something about the world, or at presenting the author's achievement. Indeed, we very often imagine that a work is one or another; this work, this atlas, or statistical work a catalog, this work of theory a reflection, this biography a narrative of achievement. We are equally unlikely, given the technologies of support for these views and those that have effaced the role of dialogue in the written work, to imagine that a work contains any dialogical elements. But we would be wrong on both counts. The written work almost inevitably contains elements of cataloging, reflection, and achievement—and it almost inevitably, too, contains elements of dialogue.

This combination of elements might be unimportant if the four views were always consistent and congruent. But the conceptions of the relationships among the geographer, her work, the research community, and society vary tremendously among written works when seen variously in these four ways. This fact is more difficult to appreciate because any work exists within a set of contexts—author, geographical community, and society—and each member of that set may emphasize and promote a different aspect of the text. The crucial matter here is this: if we look at a person's work in terms of one conception of authorship, we may see it as expressive of a particular conception of the project of geography or science; if we look at it in terms of another aspect, it may be expressive of a radically different conception.

When it comes to considering the nature of the discipline of geography, this means that individuals who in some respects appear to be alike turn out to be very different, whereas others apparently different turn out to be alike. Moreover, and somewhat more dramatically, areas of the discipline that appear from one perspective to be moving in one direction may turn out to be moving in another. To make this point clearer, I will make a few comments about David Harvey's work. Although his work is presented as that of a person who has undergone great intellectual changes, I shall argue that from another perspective it shows substantial continuity.

If we look at David Harvey's work, from *Explanation in Geography* to *The Condition of Postmodernity* to his more recent works on place and environment, we may initially be struck by the changes that have occurred, by the dramatic differences in approach.[34] In *Explanation* we find chapters on "Deductive and Inductive Forms of Reference," "Formal Statement of Theory in Geography," and the like. Four years later, in *Social Justice and the City*, there is "Revolutionary and Counter-Revolutionary Theory in Geography and the Problem of Ghetto Formation." By 1982, in *The Limits to Capital*, there is "The Contradictory Role of Ground Rent and Landed Property within the Capitalist Mode of Production." And in the recent *The Condition of Postmodernity* we find "Time-Space Compression and the Rise of Modernism as a Social Force."

From one perspective we may wish to see these changes as expressive of the rejection of one paradigm or mode of explanation and the adoption of another; conventionally this means the rejection of positivism and the adoption of Marxism. This looks like a major change. But if we look at *what* he is doing in his works, we find as well a great deal of continuity. Surely there is continuity in the ways in which the works are all, in the first instance, cast as reflections of the world. What varies is, of course, the nature of those reflections. There is, too, continuity in the ways in which his works exist as catalogs. By this I mean that from first to last the works appeal to evidence of certain—and similar—sorts. They are in a fundamental way catalogs of ideas, ideas now embodied in books, now in the built environment. And the ways in which Harvey engages in dialogue remain much the same from beginning to end; whether in *Explanation in Geography* or in *The Condition of Postmodernity*, the use of references, the tone of engagement remain the same.

Indeed, this continuity is even more clear, and more explicit, if we turn to the ways in which Harvey represents his authoring as a matter not of reflection or cataloging or dialogue but of achievement. When we look back to *Explanation*, we find the following:

> I wrote this book mainly to educate myself.... Let me explain the nature of my own ignorance as it existed prior to setting pen to paper.... [After the onset of the quantitative revolution, and] not wishing to be left behind, I naturally indulged in this fashion, but found to my consternation that I only managed to accumulate a drawer full of unpublished and unpublishable papers.... I therefore decided to devote some time to a systematic investigation of the quantitative revolution and its implications.... I then had to decide whether to abandon my philosophical attitudes (steadily accumulated from six years of indoctrination in what I

> can only call "traditional" geography at Cambridge.... I was amazed to observe how much more vigorous and vital the whole philosophy of geography became. It opened up a whole new world of thought....[35]

The theme continues in Harvey's work; he begins *Social Justice and the City* (1973) in this way:

> The biographical details of how this book came to be written are relevant to reading it since they serve to explain features in its construction that might otherwise appear peculiar. After completing a study of methodological problems in geography, which was published under the title *Explanation in Geography,* I began to explore certain philosophical issues which had deliberately been ignored in that book....
>
> The interaction between the exploration of "ideas for ideas sake" and the results of material investigation and experience provoked an evolution in my general conception of urbanism and urban problems.... The essays assembled in this volume were written at various points along an evolutionary path and therefore represent the history of an evolving viewpoint.[36]

Then, in *The Condition of Postmodernity,* we find the following:

> I cannot remember exactly when I first encountered the term post-modernism.... In recent years it has determined the standards of debate....
>
> It therefore seemed appropriate to enquire more closely into the nature of post-modernism.[37]

Each of these passages sounds very much like the following:

> [C]ertain paths that I have happened to follow ever since my youth have led me to considerations and maxims out of which I have formed a method; and this, I think, is a means to a gradual increase in my knowledge.... I shall be delighted to show in this Discourse what paths I have followed....
>
> My design, then, is not to teach here the method everyone ought to follow in order to direct his reason rightly, but only to show how I have tried to direct my own.[38]

This passage, of course, is from Descartes's *Discourse on Method* (1637). And although there is change in Harvey's work, there is real continuity in the way in which, like Descartes (or, indeed, Wittgenstein), he presents his work as a kind of philosophical autobiography, a confessional. That is to say, he presents his texts as works needing to be seen as achievement and suggests that they cannot truly be read or judged in isolation from biographical facts that characterize the nature of that achievement.

Now, there is nothing at all untoward about this; indeed, it is the rule, and in academics is even institutionalized in the form of the curriculum vitae. Rather, my point is that although one adopts one set of standards for criticizing positivist works and another for criticizing the explicit politics of the Marxist, there is another set altogether to be invoked in judging works that are confessional in nature. In part this is because a critique of a confessional is more directly and explicitly a critique of its author than is a critique of a work of positivist theory, which pretends to be authorless. And in part, it is because within this genre there is the implicit sense that each new work involves—or ought to involve—an act of transcendence, a shedding of the old self in favor of a new. The very act of writing is presented as an act of self-renewal. Whatever else we may wish to say about this way of thinking about the relationship between the author and the text, it certainly is not the conventional way of thinking about that relationship within science.

Conclusion

And so, when we look in more detail at the nature of the author of the geographical work, when we look past the simple images of reflection and cataloging, we find a much more complex landscape. First, we find the author writing works that are expressions of authorial achievement, works that show the ways in which the author has prevailed over seemingly intractable people and theories, even an intractable world. Although commonly accepted elsewhere, this view of the author and the work is seldom noted. But like the other two views, it *is* strongly supported in a range of institutions and technologies; all, in fact, are supported through conventions of style and citation.

By contrast, all of these have functioned to efface the very important ways in which a text, and any text, is at the same time dialogical, where the author is engaged in an overlapping series of dialogues with the reader, and with herself. Here a series of images of the text as permanent, external, and locationless have intersected with the institutionally entrenched codifications of the written work as property to obscure this aspect of the work and to make any work appear more univocal than it is or can be.

Indeed, the recognition of these aspects of authorship, where to write is to achieve and to engage in dialogue, in virtually every written work points to a set of related conclusions. First, the critique of a work for explicitly containing those elements must surely be greeted with suspicion, when so many works implicitly contain them. Second, the critique of a work for *not* containing them, for not being reflective or dialogical, needs equally to be greeted with suspicion. And third, the explicit attempt by an author to be reflexive, to state his or her "position," or to be dialogical, to engage in what actually *looks like* dialogue, needs to be seen as in the first instance the creation of what Foucault termed an "alter-ego," but what must surely be seen as only partial. And it will be only partial because of the inevitability imposed on the multiple nature of authorship by those technological and institutional elements that I have described.

CHAPTER SIX

The Work in the World

If we turn away from the author and to the work itself, we find that there too it is helpful to consider the ways in which the object under scrutiny is imbricated within places and sets of places. In what follows I shall look at three respects in which this is true. First, I shall consider the question of the nature of the journal and the journal article. It may seem obvious that the journal article is primarily a means of communication, but it will turn out that on the evidence provided by citation analysis, the average article does very little communicating indeed. But at the same time, the introduction of citation analysis itself is implicated in the restructuring of science and geography, where the author and the written work come to inhabit a workplace and an intellectual terrain tinged with suspicion.

Similarly, it is commonplace to imagine that the primary way in which a newly emerging subdiscipline or approach gains credence is through force of argument, or perhaps weight of empirical evidence. Here even the most cynical view, where publications are seen as ways in which friends publish and cite one another's work, often supports this same sense of the written work as a container of facts and concepts. But if we look at a series of programmatic works within the discipline, we find little reason to believe that the enduring popularity of one over another results from its having greater theoretical or empirical force. Rather, a fundamental reason for success appears to be the way in which certain sorts of programmatic works offer the promise of a newly organized place of research, which brings together ideas, people, techniques, and technologies, all of which appear likely to benefit from adherence to the ideas spelled out.

At the same time, and finally, there is an important sense in which this establishment of broader systems of alliances, ones with institutions outside of the academy, is itself associated with a refiguration of what it means to be a work and an author, and with the ways in which the products of geography and science can be seen as fitting within a traditional system of rights and responsibilities. Although this refiguration has happened in somewhat different ways elsewhere, in physics and biology, for example, we shall find that the development of geographic information systems as means for analysis and representation is having a fundamental impact on the question of what it means to be a geographer.

Writing without Representing: The Journal in an Era of Reward and Surveillance

I spoke earlier about the process of citation and about the ways in which the establishment of systems for the analysis of citations has been associated with particular ways of looking at the author and the written work. But the use of citation analysis has yielded an unexpected conclusion. Indeed, an analysis of the citation literature in the social sciences can lead to only one conclusion: the journal is simply not the primary medium of communication in the social sciences. Despite the prestige that journals have, few academics read them, few cite the articles that they read, and few write for them. According to Fritz Machlup and Kenneth Leeson's survey of economists, sixty-four percent read no articles in a year (out of seventy-five major articles in eight journals) with "great care."[1] On the basis of their data it does not seem unreasonable to suggest that the average attendee of an annual meeting of the Institute of British Geographers (IBG) or Association of American Geographers (AAG) hears more papers there than he or she reads in a year. Indeed, as early as 1970 researchers found in a broad study (which included geography) that the great majority of academic papers that ultimately were published in academic journals had previously appeared elsewhere—as presentations at professional meetings, or as proceedings, preprints, or monographs—and that many were later republished.[2] Many who had read or heard the originals were unaware that they had later been published in peer-reviewed journals.

Further, in science as a whole, about one-half of all journal articles are never cited; of those that are, few are cited very often; and of those, most are cited soon after publication and then rarely again. In geography J. W. R. Whitehand reported that about seventy percent of works were never cited.[3] By contrast, "three authors (1.5 per cent) accounted for about one-half of the citations and 12 per cent of authors accounted for 85 per cent of citations."[4] There were only thirty-two geographers whose works were cited more than one hundred times between 1971 and 1975; of those, all but nine received fewer than two hundred citations. Finally, he reported that the average member of the AAG published 0.15 papers in 1982, or one every six or seven years. More recently, he noted that in 1987 there were a total of two hundred citations of articles published in the (relatively) highly cited *Progress in Human Geography,* and that of those, one-half were citations of articles published in the most recent four years, and one-half were of earlier articles. Hence, few scholars read very many articles, few write articles, and few cite articles.[5]

If journals are not the primary means of academic communication, two questions arise. First, what is that medium? And second, what functions do journals fulfill? I shall answer the first question in only the sketchiest way. It seems clear that the development, particularly in the United States, of a large discipline of geography has led to a sort of fracturing. Despite their size, national meetings contribute to that fracturing to the extent that they require that a person choose from among conflicting paper sessions. Further, more and more work is done in smaller groups. There are a number of such groups, among them specialty groups and ad hoc groupings of scholars of, for example, industrial restructuring, culture, or gender, who meet together. Finally, cheaper means of communication and transportation have eased the transition to this form of organization; in an important sense the development of xerography and computer networks has meant a renascence of the erudite letter.[6]

The second question, about the functions that journals do fulfill, is more immediately to the point here. I would argue that journals today increasingly, and alongside their traditional function of communication, fulfill three major functions. First, and this is not new, they are archives, which function to establish intellectual property rights. Second, and I have discussed this earlier, they are means for the establishment of discipline. And third, they are part of a system of surveillance. This is increasingly an important function and deserves more attention.

The development of surveillance arises as the development of citation indexes and analysis has in a sense split the operation of the reward system into two parts. The traditional system remains, but through the development of more informal modes of organization, its functioning has changed; it now operates at different scales than previously. But that system has been overlaid by a kind of "shadow" system, the system of surveillance. It has, of course, been typical of scientists and other professionals to argue that they possess a special knowledge, which renders them preeminently and uniquely qualified to judge the work of their peers. But citation indexing systems move the gathering of data out of the academy and into the marketplace, while they suggest that analysis of those data is simple enough that anyone can do it. In that sense they allow for the establishment of a second system, of surveillance and control, that is outside of science's traditional reward system, and thereby outside of the control of scientists themselves.

In fact, the lack of control and the acquiescence of geographers to it have both been remarkably complete. In a 1984 article, Anthony C. Gatrell and Anthony Smith noted that several of their list of the "top" twenty-two general and human geography journals are not covered by the Institute for Scientific Information (ISI).[7] They left the matter at that, and they failed to follow up. In fact, ISI summarily, and without explanation, denies requests by journals to be included. Indeed, it is hard to avoid the conclusion that those decisions are made, fundamentally, with an attention to financial considerations. As ISI's "coverage specialist" noted in response to a request made in 1987 by Basil Blackwell for the inclusion of *Antipode* in the *Social Science Citation Index* and in *Current Contents*, "Our available space in the *SSCI* and *CC/S&BS* is very limited." Further communications from ISI were even less informative: "We completed our evaluation [of *Antipode* and the *LSE Quarterly*] earlier in the year and decided not to include it," and then, "We are not currently covering *Antipode* in any of our products. If I can be of assistance in any other matters, please do not hesitate to contact me again."[8]

Although the ISI has an advisory board, on which serve several well-known sociologists of science, like Robert K. Merton and Harriet Zuckerman, the board contains no geographers. Hence, geographers, here, have decided by default to allow a group of non-geographers to determine which of their journals are important. Because some authors consider it important that their works be indexed, this decision means that authors will be less inclined to submit their articles to unindexed journals, which will, thereby, become less influential in a self-fulfilling cycle.

The fact that there are now two systems is in some respects masked by the scientific trappings adopted within the surveillance system. Yet what fundamentally distinguishes it from the traditional reward system is that it operates within the context of different goals and norms. Indeed, with both the control of data about "quality" in science and the analysis of those data now routinely in the hands of non-scientists, it is possible through the administration of rewards to manipulate the system of science in highly politicized ways.

It is ironic, but not surprising, that citation indexes, the very tools that have been so effective at indicating the lack of effectiveness of journals as means of communication, have been equally successful at presenting the analysis of those journals as a better means of surveillance. The success of the indexes is, in part, related to their use of technology; they seem thereby to have adopted the logic of science. In doing so they appear to have laid the groundwork for a more systematic means of comparing works, and thus finding ethical misdeeds.

But at the same time, this use of technology seems to be an example of what Langdon Winner termed "autonomous technology," technology out of control.[9] And to the extent that it tends to move the locus of control of science from practicing scientists, it increases the sense of heteronomy, the sense that things are run "from outside" and that the scientist or geographer is merely a cog in a wheel—and not a moral agent.

So the rise of a formal system of citations and their analysis is inextricably related to the issue of ethics, and it is thereby related to what is sometimes termed "self-plagiarism."[10] Self-plagiarism is the process of publishing the same or similar articles in several places, and where one's output is measured only in terms of titles and citations, it becomes easy to manipulate the system. The issue of self-plagiarism is, of course, fundamentally related to the question of the ownership of ideas. From one point of view the matter is simple; most reuse of one's work is wrong because it violates provisions of the copyright law, which prohibits the republication without permission of works that have previously been copyrighted. Yet even there things are not entirely clear, because the copyright law applies in only the most contorted way in academic publishing. In crafting an article an author goes to considerable expense but then gives it gratis (in some cases an author is required to pay "page charges") to a journal, which may charge several thousand dollars per volume, which in turn charges him or her for copies of the article, then charges students for the rights to copies of it, and may then charge the original author for publishing parts of it elsewhere. In fact, given the way in which the understanding of copyright by academic publishers caricatures normal notions of rights and ownership, Gillian Page,

Robert Campbell, and Jack Meadows suggest that the law, as applied to journal publishing, would probably not stand up in court, particularly since the author receives nothing in exchange for value rendered.[11] Hence, it is of little surprise that academics fail to take the copyright law seriously, and regularly and systematically flout it in their teaching.

Leaving aside the legal issue, it should be clear that as contentious an ethical issue as self-plagiarism will arise in very different ways—or not at all—in different parts of the field and at different times. In some fields, where written works are more matters of gradual achievement than of the cataloging or reflecting of the world, simple works will more often appear questionable. But in either case the possibility of accurately finding cases of self-plagiarism depends, in addition, on the systematization of writing; here citations and their indexes can be especially important.

Unfortunately, as early as 1965, Norman Kaplan noted:

> My very own preliminary studies lead me to suspect that the citation practices of scientists today are in large part a social device for coping with problems of property rights and priority claims. Only incidentally do these citations serve as a careful and accurate reconstruction of the scholarly precursors of one's own contribution.[12]

In fact, sociologists of science have virtually despaired of the possibility of adequately conceptualizing the actual workings of the citation process; the best that many have been able to come up with is that learning to make citations occurs "osmotically," through a process of absorption.

Indeed, if the "writing up" of research results is not rhetorically neutral, neither is the citation process simple or systematic. For example, Michael Moravcsik and Poovanalingam Murugesan point to several types of references.[13] They may be conceptual or operational, organic or perfunctory, evolutionary or juxtapositional, or confirmational or negational. Somewhat more cynically, Frederick Thorne lists a very different set of types of citation, including hat-tipping (showing that one has done his or her homework), over-detailed (citing everything possible), self-serving, premeditated (choosing references that the editor will like), projective (citing only members of a particular school), conspiratorial ("I'll cite you if you cite me"), obsolete, and pandering (citing useless work because readers demand it).[14] Whereas Moravcsik and Murugesan and Thorne provide reasons for believing that the process is not simple, J. B. Bavelas points at a more fundamental set of difficulties.[15] He notes that a basic reason for citing particular works is to show that we "know the literature," and notes that because this is an appeal to what members of the profession consider important, citing is really an appeal to consensus, and thus reveals more about the status of the discipline than about the intellectual origins of ideas. Hence, the same author may in writing for different audiences cite different sources for the same ideas, merely as a means of making clear to that audience that he or she knows the literature.

Perhaps more fundamentally, though, the use of citation indexes as measures of quality rests on a particular and limited view of science. If we see the sciences as fundamentally repositories for information and tools by which to manipulate that information, then it might well make sense to believe that a key finding could and should be cited. But if we turn to certain of the social sciences, we find a fundamental difficulty, one that Arthur Danto some years ago elucidated through the image of the ideal chronicle, an image that, ironically, is strongly reminiscent of Eugene Garfield's World Brain.[16] Imagine, he argued, that we have an "ideal" chronicle, a complete accounting of every event that occurs. What sort of statements would it be impossible to make? One would, it turns out, be unable to make statements like "World War II began today" or "The seeds of the restructuring of the Western industrial economy were planted today," because those accounts are necessarily retrospective. Indeed, much of what occurs in the social sciences is a matter of trying to understand what has gone before, and the questions that are at one time seen as important are later seen as less important. Hence, within the social sciences, works, whether they are reflections, catalogs, achievements, or dialogues, have a built-in obsolescence, based less on the ease with which theoretical constructs are criticized than on the changing context from which questions are asked. For this reason a reward system built upon the analysis of citations must fundamentally exhibit biases toward certain types of literature; it is no accident that the most cited works in the sciences are descriptions of laboratory procedures and that the most cited article in *Progress in Human Geography* is Andrew Sayer's "Explanation in Human Geography," an essay that is often used as a totem by geographers claiming to be "realists."[17] Thus, the citation and publication systems are not neutral evidence that can be used to judge individuals, departments, universities, and disciplines; rather, they are in a fundamental sense reflections of the power relations that the various individuals involved perceive.

Garfield and his employees at ISI have done their best to convince skeptics that these are not substantial problems.[18] But it would be unfair to suggest that it was these ministrations that helped legitimate citation analysis; indeed, because of its technical structure, the citation index seemed to many to be beyond question, and where—again and again—critics attacked it, they almost inevitably accorded it some degree of legitimacy. This has certainly been the case in geography, where we find in David Lee and Arthur Evans, B. L. Turner and William Meyer, J. W. R. Whitehand, and Neil Wrigley and Stephen Matthews lip service to the problems involved, and then the use of those data as though the problems simply do not exist.[19] In doing so they privilege the notion of "information." They seem often to embrace a form of the naturalistic fallacy as they operationalize "truth" in terms of "citedness." And they run the constant risk that they, their readers, or the less sophisticated users of their techniques will render all that is included in the indexes, whether good, bad, or preposterous, equal.

And in doing so they suggest a reductive, and because of the biases of citation patterns, misleading means for assessing similarities and differences among works. Here, as with the case of abstracts, the means for avoiding charges of self-plagiarism are readily available to those whose works offer the possibility of varying citations. But those who produce works with similar patterns of citations will, even if the works themselves are very different, be under a cloud of suspicion.

In fact, though, the problem of assessing written works, and of finding cases of self-plagiarism, is even more difficult than this indicates. In some subdisciplines, ambiguity and vagueness appear to be inevitable accompaniments to the process of writing and research; this is particularly true in those areas where the development of standards of practice has not been accompanied by the development of greater control over the objects of research.[20] Nowhere is this more true than in the social sciences; as Harold Garfinkel, Michael Lynch, and Eric Livingston have described them, they are "talking sciences" that "achieve in texts, not elsewhere, the observability and practical objectivity of their phenomena."[21]

Hence, the nature of the text, and especially where the text is an achievement, has implications for the issue of self-plagiarism. Although some appear to believe that this process is both easily discovered and clearly wrong, in a "talking science" matters may not be quite so clear. From the perspective of someone doing cultural, philosophical, or historical analysis, the world is likely seen as a complex, ambiguous, and richly textured place, about which there are few certainties. Hence, a scholar working within one of those traditions is likely to spend long periods working on a very few projects, and to spend a great deal of time "tuning" results tentative first in one way, then in another. In that process there may very well be themes and problems that recur. For someone interested in, say, phenomenology, the relationship between phenomenology in geography and the traditional works of Edmund Husserl may be a recurrent issue; having once laid out in writing that relationship, the author may be tempted to reuse that text in different contexts and for different reasons in later works. Why, after all, ought one to rewrite something merely for the sake of rewriting it? And why, especially when in different contexts a text may have different rhetorical forces, and hence different meanings.

In fact, having developed a comfortable relationship with such issues, an author may well unwittingly paraphrase an earlier text. Is this self-plagiarism? It appears to be and may, too, be legally wrong. Is it morally wrong? This is less clear. What if an author says "exactly" the same thing in completely different terms? If in supporting different arguments in different papers I quote the same source material, is this self-plagiarism? What if in two papers, each with a different point, I quote *only* the same materials? What if the quotations make up the majority of the paper? What if I write the same material for radically different audiences? What if in different papers I use not the same English phrases, but rather the same mathematical or logical transformations? There presumably is no problem if, like addition or *modus ponens,* they are in the public domain. But what if they are my inventions? I may fail to cite their sources, believing them now standard, but what if an editor considers them the most interesting part of the paper? Does this mean that I have plagiarized my own work?

These are difficult questions and surely lack unambiguous answers. But the matter is actually even more confusing. If in two different papers I ask, note as important, but do not answer the same set of *questions,* is that self-plagiarism? Once one has allowed that there is *something* called self-plagiarism, it becomes difficult to answer these and similar questions without making a series of assumptions, including ones about the mental states of the author in question, that are highly questionable. At the same time, one appears led to the oddest of conclusions, because one seems driven to conclude that much of the most compelling work from later in the careers of some of the giants of Western thought, Plato and Aristotle, Kant, Hegel, Marx, and Nietzsche, ought never to have been published. To adopt a strict definition of self-plagiarism is to aver that there is no place for the gradual, lifelong working out of a set of problems and questions, or for the analysis of those workings.

And so, the development of citation indexes, which appear to be neutral and scientific means of determining the structure of a discipline, in fact allows for the use of a system of surveillance that may itself be detrimental to the achievement of the goals that many in science profess, and in several ways. It both allows people to engage in a sort of work that many do not see as legitimate and refuses them the possibility of engaging in a sort of work that many would value. In the end, the adoption of such systems points to a world in which the written work, far from being simply the final representation of a search for the truth, is fundamentally an element in a system of surveillance, and to a workplace in which this function is embraced, both by those who wish to use it for surveillance and, in a Foucauldian twist, by those who are its victims.

Manufacturing Alliances and the Roads to Success

It is common to imagine that the way in which a work succeeds is through its somehow being "right." A work that expresses the truth will, barring accidents, be seen as truthful and will be accepted into the body of knowledge of the discipline. As we have seen in the previous section, when we concern ourselves with a single author, this appears, often enough, not to be exactly the case. But if we now turn to a certain type of work, we find that there, too, the reality is different from the image.

I have in mind here what I term "programmatic statements," those written works that attempt to convert a reading public from one point of view to another. I will begin by looking at a famous paper by John Q. Stewart and William Warntz.[22] My strategy will be comparative; I shall look at their work in the context of others like it and consider the adequacy of the means that they use in comparison with those used by their peers.

The first question is, of course, simple: What *is* their peer group? The answer may seem obvious; it consists of those geographers who have been involved in the project of mathematical modeling and theorizing about mathematical models, those who worry that journals like *Geographical Analysis* may be becoming too historical when they sometimes publish articles with bibliographical references.

But these geographers, I would argue, are not Stewart and Warntz's peers. Rather, their peers consist of another group entirely. They include Anne Buttimer, Michael Dear, Leonard Guelke, Allen Pred, Andrew Sayer, David Harvey, and Ted Relph.[23] Now this may seem a surprising group. But what they all have in common is this: they have published well-known articles of a particular form, typically in widely read journals. I shall call these articles "programmatic statements." These are articles that attempt to recast geographical theory but also to go farther, to a recasting of the discipline of geography itself. I shall argue that when we consider Stewart and Warntz's article as a member of this group, it turns out not to be terribly sophisticated from a theoretical point of view, but nonetheless to be far more skillfully wrought than the others that I have mentioned. It is skillful in its strategies of exclusion and appropriation, in its placing some geographers at the margin and placing itself at a new frontier.

The traditional way of thinking about the history of geography, or, indeed, of any discipline, would not, actually, make the distinction that I have just made, because it would see—at least in the long term—the development of any discipline as "going with" the development of a set of ideas. Bad ideas and theories, on this view, drop out; good ones come along and replace them. Yet, to be quite honest, if we read the works that I have just mentioned retrospectively, this development does not appear to be what has happened. Well-written and thought-out works advocating approaches as diverse as phenomenology and time-geography appear to have fallen on deaf ears, while geographers were hanging off the sides of Stewart and Warntz's jury-rigged bandwagon. So this evolutionary approach just will not do as a way of thinking about the development of the discipline.

One might then want to refer here to the literature of rhetoric and argue that Stewart and Warntz's article needs to be analyzed not simply as a piece of scientific argument but rather *as* one of rhetoric. But here a different and, I think, preferable way of looking at scientific texts has recently emerged, in the work of people like Michel Callon, Bruno Latour, and John Law.[24] It seems to me that their work has much to offer in the way of understanding just what Stewart and Warntz were doing. Although I have a series of reservations about some features of their work, based in part on the ways in which they deal with the issue of relativism, I shall draw on it in attempting to understand Stewart and Warntz's essay and other programmatic statements.

ON PRESSING PROGRAMS AND CREATING ALLIES

I shall begin not with Stewart and Warntz but rather with their peers; in the interests of brevity I shall leave aside works by Harvey and Sayer, as well as similar ones on Q-analysis, localities, and so on. I shall move through the remainder in a rough order, from those least to those most like Stewart and Warntz, and in that way shall highlight some features that I think make each work interesting. I shall begin with Leonard Guelke's paper, "An Idealist Alternative in Human Geography" (1974). Guelke for a time was pouring out these articles; there are perhaps six or eight, all with similar themes. He begins this one with these assertions:

> The search for theory has become a hallmark of modern human geography.... Many... cannot agree on the value of the theory thus far developed.... Positivism assigns disciplines without theory a low prescientific rating.... On the idealist philosophy of explanation, which is advocated here, the need for theory in human geography is denied....[25]

So within the first paragraph the game is on. There is something called "positivism," which is evidently bad and is attempting to call the shots within geography. It demands that geographers create theories, even though theories appear to be of limited use. We need an alternative, and something called "idealism" may be it.

Guelke's next step is to define the "core of the idealist position," which is that "the explanation of rational human behavior demands a mode of understanding quite different from that which is appropriate to nonrational human and natural phenomena" (193). He then moves into the exposition of his position, which is derived, he says, from R. G. Collingwood's *The Idea of History*.[26] It is concerned with people as rational beings, who have intentions that, in turn, direct their actions. Indeed, "the explanation of an action is complete when the agent's goal and theoretical understanding of his situation have been discovered" (197). This involves a "rethinking" wherein one "seeks to discover the way in which a geographical agent construed his situation in order to see the link between thought and action" (198).

Although Guelke has at the outset asserted a divide between his own project and that of "positivism," he now suggests that "there are a number of similarities between scientific theories and rational interpretations with regard to criteria of acceptability" (201). Finally, he closes, as he began, with these assertions:

> The idea that human geographers ought to attempt to emulate physical scientists in search of theory overlooks the fact that man himself is a theoretical animal whose actions are based on the theoretical understanding of his situation.... This approach is no more subjective than that employed by positivists.... The idealist human geographer aims at providing a true explanation of the situation he investigates. (202)

Some of the problems with this article, the reasons for its failure, will become clear as I comment on others. I would, though, note the following. First, the article is an essay with the following form: A force is pressing on geography from without. This force is confused, but geographers are being taken in by it. Instead, we need to reject it. If we do so, by adopting the right approach, we will prevail. Indeed, the right approach can do all that the outside force claims to be able to do but in fact cannot. This turns out to be a form that repeats itself with great regularity in programmatic works. And second, Guelke's article is remarkably devoid of references. One finds Collingwood, Evans-Pritchard, Popper, Walsh, and Hempel. Given that the article is 7,500 words long, Guelke shows what by today's standards would be remarkable restraint in citation.

If we turn to a second essay, Anne Buttimer's "Grasping the Dynamism of the Lifeworld" (1976), we find an argument structured in much the same way. It begins this way:

> Recent attempts by geographers to explore the human experience of space have focussed on overt behavior and its cognitive foundations. The language and style of our descriptions, however, often fail to speak in categories appropriate for the elucidation of lived experience. . . . Phenomenologists provide some guidelines for this task.[27]

Like Guelke, Buttimer turns immediately to a critique. We are in a "malaise," a "rut." The explanations of social scientists are "opaque and static." They merely "record." Hence, philosophers and geographers face a "common task," to "reconcile heart and mind, knowledge and action." "Phenomenologists have been the most articulate spokesmen for this endeavor" (278).

From there she moves, again like Guelke, to an exposition of her position. But here, she notes, problems immediately arise:

> Phenomenology is not easy to define. . . . There are difficulties . . . in relating the notion of "lived world" to geographic language and endeavor. . . . It is in the spirit of the phenomenological purpose, then, rather than in the practice of phenomenological procedures, that one finds direction. (279–80)

Buttimer next turns to "the human experience of space," to lived and representational space, ways of knowing, experience, the sense of place, social space, and time space rhythms and milieu. Then, considering possible areas of interaction between geographers and philosophers, she takes on a confessional tone: "Personal experience has shown me . . . ," "I have realized now . . . ," and "My own reflections have highlighted . . ." (289). She concludes that we need the help of phenomenology to

> move us toward a keener sense of self-knowledge and identity; it will create a thirst for wholeness in experience and a transcendence of a priori categories in research. . . . It could also help us transcend the artificial barriers which our Western intellectual heritage has placed between mind and being, between intellectual and moral, the true and the good in our life worlds. (292)

Now, in one sense Buttimer's work reads very much like Guelke's. The two works are similarly structured, and both present views that they take to be real alternatives to some other approach to geography. And both present views that focus on human experience and motivation. Yet, in another sense they are strikingly different. Buttimer's essay is extensively footnoted. Beginning with Heidegger, she moves quickly among geography, philosophy, anthropology, and psychology. Her references to geography themselves cover a broad spectrum, from the then new humanistic geography, to more theoretical works, to those of Marxists and time geographers. Whereas Guelke's position was presented as a whole, with an almost "take it or leave it" attitude, Buttimer's is presented as far more dynamic, as still evolving, and evolving out of the interaction of a wider range of actors.

This position actually turns out to be a disadvantage to the extent that it arises from the fact that, as she put it, "the still unanswered questions about the relationship between phenomenology and geography are many and complex" (290). Her position means that she leaves matters quite open-ended; whereas in one sense she is attempting to convert geographers to a new way of working, in another she is providing them with a loose set of concepts and tools and simply letting them go. She has left no real way to know whether people actually are converts. And so, today among geographers who term themselves phenomenologists, there are people as diverse as David Seamon and John Pickles, whose works are very different one from another and who appear not to converse.

Some of this difficulty is removed in Allen Pred's "The Choreography of Existence: Comments on Hägerstrand's Time-Geography and Its Usefulness" (1977). Pred offers three ways of avoiding this difficulty and of establishing a means for the defining of converts, and hence for success. First, he attempts to remove uncertainty by directing the attention of geographers away from other disciplines to their own. "The need," he begins in his abstract, "for human geography to turn inward with respect to the definition of research problems and the use of structures is pointed to. . . ."[28] Second, although his essay is in some ways structured like those of Guelke and Buttimer, he immediately sets off to describe "planning applications," "applications related to traditional research themes in human geography," and "other possible uses." Each item is quite brief, and they are presented with almost a staccato rhythm: Time-geography can assess regional variations in accessibility to medical care, examine conflicts, examine agricultural transformation, show the effects of possible changes in the work week, formalize historical geography, show impacts of technical innovation, understand innovation in the arts, describe migration, and so on, an almost breathtaking array of possible uses.

Although Buttimer appears superficially to have used a similar strategy, there is a significant difference between hers and Pred's. Pred suggests a diverse group of uses but does so in a way that might mobilize a substantially larger and "deeper" set of allies. He suggests practical applications of his approach and ones that appear to be immediately useful. These might, at least in principle, result in the rapid development of groups of professionals using his techniques, thereby feeding back into the academic establishment.

And this strategy leads to the third; time-geography is presented as resting on a limited and simple set of concepts. As Pred put it, it is "disarmingly simple in composition and ambitious in design."[29] The concepts are so simple that they appear not to require a set of intermediaries to translate them into terms that the average person can comprehend; they can be presented in a simple, straightforward way, often with the aid of visual representations.

The failure of time-geography to attain wide acceptance among academic geographers strikes me as having derived from its having several of the same features that earlier doomed the study of mental maps. Both seemed at the outset to be fertile metaphors, but geographers came quickly to believe that their fertility had been exhausted. Still, this failure ought not to blind us to the sophistication with which they were presented.

Much of that same sophistication can be found, some may be surprised to see, in Michael Dear's "The Postmodern Challenge: Reconstructing Human Geography" (1988). Granted, an essay that begins with a listing of AAG specialty groups and then moves on to a discussion of a Steve's ice cream advertisement may not seem to be a likely candidate for an award for sophistication of presentation. But there is more to it than that.

The essay starts in a now familiar way:

> This is a time of intellectual crisis . . . [but also of] remarkable opportunity. . . . By confronting the intellectual disarray in human geography, and addressing the challenge of postmodernism and deconstruction, the discipline could attain a pivotal role in the social sciences and humanities.[30]

Dear goes on to describe the nature of the crisis within geography, where at once the discipline appears to be becoming increasingly diverse, even splintered (hence the list of specialty groups), and to be subject to a kind of siege mentality, as it undergoes criticism from those outside. He then describes the nature of two intellectual movements—postmodernism and deconstruction. Here he suggests that we need to reconstruct human geography:

> I shall recast the concerns of human geography within the context of mainstream social theory, and thereby demonstrate (1) geography's pivotal role in a reconstituted social science; (2) a plausible restructuring of the internal order of the discipline; and (3) the insight and advancement made possible through embracing postmodernism.[31]

He concludes that all of this will involve a dramatic redefinition of the subdisciplines within geography, which will place the economic, the political, and the social at the fore. According to Dear, those who have made their homes in "microcomputers," "audio-visual techniques," and "librarianship" had best start retooling at the first opportunity.

Now in one sense this whole project seems quite far-fetched. And yet, Dear has done several things that are quite similar to those that Pred has done. Perhaps most important, he has offered a version of geography within which a set of people now marginal, those doing certain types of economic, social, and political geography, can come to the fore of the discipline; that is, he has attempted to create a strong set of allies. Second, he has attempted to create a well-defined intellectual core, to bind this group together and coordinate their future work. And third, he has done so in a way that he sees as establishing alliances between geographers and non-geographers; here the hope is that the result will be the movement of the discipline to the center of social theory, and hence into the center of vision of those with institutional power, and that the consequence will be more resources for the discipline. In this way, too, he has suggested a strategy for the resuscitation of what he began by asserting to be an almost moribund discipline; in this way he hopes to create a further set of allies within the field, who would otherwise be utterly uninterested in postmodernism and its associated doctrines and methods.

Eight years after Dear's essay, it now appears clear that Dear's dream is not to be realized. There are a number of reasons for its failure, but here I shall point only to two. First, Dear badly misjudged the importance of what he termed "social theory." A quick look at university course syllabi will show that far from being central to the practice of the social sciences, social theory is itself quite marginal. And second, he failed to present a project that had any obvious allies outside of the academy. In both ways he limited the possible appeal of his program.

STEWART, WARNTZ, AND THE CREATION OF A GEOGRAPHICAL INSTITUTION

Turning now to Stewart and Warntz, we find what turns out to have been a remarkably wide-ranging presentation of a program. It is a presentation that succeeds in the way that Pred's and Dear's do while avoiding their mistakes.

Like the others whom I have considered, Stewart and Warntz begin with a critique:

> Those who insist that in considering the phenomena of human geography we should "let the data suggest their own terms of study" have done much to impede the development of the subject.... But general regularities and systematic relationships are not discovered by such an examination of separated localities.[32]

They characterize this mainstream approach as the "microscopic," and argue that we need, instead, to develop a macroscopic approach. Such an approach, which can provide a "sufficiently abstract and subtle measure of position... has been published," (168) they continue; it is the approach of social physics.

At this point they launch into a set of equations; invoking "Newton's law of gravitation," they offer a series of concepts, with "potential of population" the most prominent; here "the number of people replaces mass throughout" (171). We find, in short order, that

> Rural population density (persons per sq. mi.) = .0336 V^2
> Rural nonfarm pop. density = .000562 V^3
> Miles of rural free delivery routes per sq. mile = .00517 $V^{3/2}$. (172)

There are additional findings:

> Log income density = –4.92 + 2.58 log U (r = .87)
> Log telephone wire density = –5.70 + 2.81 log U (r = .85). (176)

And so on.

Next, Stewart and Warntz move into an aside on coefficients of correlation, then back into more findings, now about average county size, farm size, Federal Reserve District size, and on and on. They suggest that the results that they have obtained are there on the surface, waiting to be discovered. "The map shows" this; "Inspection of the map... strongly suggests" this; "This same systematic relationship can be observed"; "The value of land... likewise showed" (178–79).

Finally, they conclude,

> Geographers have the opportunity to make their contributions to integrated social science especially through studying distances not only as a social factor but as a factor in macroscopic geographic equilibria. If geographers avail themselves of this opportunity, the subject will once more progress. If they again fail, it will remain out of the mainstream of current high-level academic and philosophic thought.
>
> Let the geographers who hesitate to exploit the enormous advantages presented by the concept of potential of population be reminded that "population density," a concept now unquestioningly accepted and employed by geographers, originated in physics. . . .
>
> The geographical profession at large was rather hesitant . . . [but] today the concept of population density is readily accepted. (181–83)

Remarkably, with this, and then a brief poem, they end their essay.

Now, on the basis of what I have said, this may not seem to be a very sophisticated or appealing essay, and one may wonder why I see it as in its way quite successful. I would argue that it is successful for two reasons. First, it establishes a group of allies that is larger, more varied, and more substantial than that established by any of the other authors whom I have mentioned. And second, it constricts the amount of room they have in which to maneuver. I will say a little about each feature.

We recall that Guelke attempted to ally geography with history, and Buttimer attempted to establish a wider but purely academic set of allies, whereas Pred moved beyond the academy into the planning profession. Stewart and Warntz make this same external move, and more. They provide a few citations to geography journals but more to economics, sociology, and regional science. They create another sort of alliance through their use of statistics; their short asides are structured to leave the impression that there is much more to be learned. And they create another alliance through their claims of obedience to physics.

But this is not all. We note in their acknowledgments a further sort of alliance, to the American Geographical Society, Princeton University, the Ford Foundation, the Institute for Advanced Study, the Research Corporation, and the Rockefeller Foundation. Here, of course, the suggestion is that this sort of research involves substantial links with groups not traditionally associated with geography.

At several points the authors thank graduate students for performing calculations; in this way they create a different set of alliances, as they promote the notion of research as teamwork, and not the solitary activity that we saw in Guelke and Buttimer.

Their use of cartography—again with the requisite acknowledgments—offers an additional sort of connection. Moreover, one map was "prepared using IBM equipment, including sorter, tabulator, and summary punch. The isopotential lines were drawn on the basis of logical contouring from the computed values for 115 control points" (174). And finally, most of the data that they use were provided by the U.S. government.

All of these alliances, I would argue, have historically been far more important to Stewart and Warntz's project than have been the niceties of theory. What they have done, in stark contrast to Guelke and Buttimer, and even to Pred and Dear, is create an image of a geography that has a substantially larger group of players, of people with an interest in its success. The discipline is no longer seen simply as one in which a single individual or small group go into the field and observe, then go home and describe. Rather, this seasonal ebb and flow is replaced by a more constant process of inputs and outputs, with inputs of time and money from a wider range of institutions, with a wider group—including data collectors and computer manufacturers—ready to offer to take that money. Here traditional maps and photographs have been dismissed as merely illustrative; in the new macroscopic geography they will truly be means for analysis, and hence will be in greater—and constant—demand. All of this, in turn, means a new market for technical support staff both within the discipline and outside of it. We have begun to see the effects of this as the advertisements in trade magazines for geographic information systems and the displays at professional meetings begin to be as impressive as those that for years have graced the pages of *Science.*

This advanced technology would be of limited use in the absence of a second feature of Stewart and Warntz's work, and that concerns the limitations that they place on the kinds of work that can be done. If we look again to Guelke and Buttimer, and also to Dear, we find that their proposals provide little in the way of guidance with respect to the actual carrying out of research. Certainly almost anyone can claim to be postmodern, just as anyone can claim to be doing phenomenology. Now, in one sense, of course, the social physics movement has been sidetracked and currently exists in a form different from that which was once envisaged. In fact, the hoped-for human laws failed to appear. And yet in another sense, it has succeeded. It has done so to the extent that it has provided an image of the task of geography wherein the subjects of study—people—are unambiguously domesticated. There can be no human subject more docile than one of Stewart and Warntz's social atoms, and we find them today in the pages of a wide range of geographical journals.

Alongside the docility of the human subject lies an image of research wherein the concern with detail must always be subordinated to the concern for an underlying pattern, which is assumed to exist. Where the authors point to one equation after another and note that in each case the correlation coefficient is the same, they are pointing to a human world within which a high degree of order exists and need only be found.

In these two ways, in the creation of a vast network of potential alliances and in the establishment of formal and easily comprehended restrictions on the nature of geographical research, Stewart and Warntz provided a new vision of the discipline. It was a vision in which the subtleties of conceptual analysis were left behind, just as was work that required the certification of the worker through, for example, shared field experiences. Indeed, I would argue that their work shows less the influence of science than the influence of a broader climate of opinion in which science is lionized. But in retrospect, and especially as we consider the fates of the various alternatives that I have considered, it becomes clear that these conceptual and philosophical issues have had far less impact than has the establishment of the new set of institutional structures that they invoke.

It might be argued that I am giving Stewart and Warntz too much credit. And certainly, a quick reading of their essay shows that it lacks any of the literary finesse that we find in Buttimer, Pred, and Dear. Neither do we find much in the way of explicit self-consciousness. One wonders whether they really thought that an argument like the following was at all coherent:

> Physicists invented density maps.
> Density maps are useful in geography.
> Physicists invented potentials.
> Therefore, potentials must be useful in geography.

But I would suggest that the features of their essay and the others that I have considered need not have been inserted into the works with the effects that I have mentioned in mind. Rather, I would suggest that Stewart and Warntz speak of IBM, physicists, correlation coefficients, the Rockefeller Foundation, potentials, economists, and logarithms not so much because they are intending to use them to rhetorical or political ends but rather because these players—Latour calls them "actors"—constitute a natural part of their present and hoped-for institutional landscape. Hence, it makes more sense to see Stewart and Warntz not as politically savvy individuals intent on redefining the realm of geography but rather as two individuals who have internalized a particular climate of opinion and are merely its passive voice. I would argue, in fact, that it is this attention to the place of the work in the world that makes their work stand out from even the best of other programmatic statements.

Rethinking the Place of the Written Work: The Case of Geographic Information Systems

The evolution of the discipline of geography has surely followed a path that Stewart and Warntz could not have predicted. But nonetheless, this evolution has happened in a way that is closely tied to the processes of the creation of alliances. And nowhere is this process more obvious than in the matter of geographic information systems. Developed over the last thirty or so years, this marriage of the computer, the map, and the quantitative revolution has at its core been one of the creation of such alliances. Indeed, in geographic information systems the creation of these alliances has advanced to the point that we can begin to see a refiguration of a major part of the discipline, a refiguration around the production of a new version of the written work. Central to this refiguration has been a rethinking of the very questions of what it is to do geography, or science, and of the structure of academic practice.

Within the scientific academy, as well as outside, it has long been imagined that in any scientific project one can distinguish between some individual or group that constitutes a creative and thinking head, and other individuals and groups that can be considered merely as laboring bodies. This image, so central to the division of labor, has long been appealed to as a means for guiding discourse about rights and responsibilities, with different ones accruing to the head and the body. But the sets of practices that have developed around the mix of hardware, software, and data within a geographic information system have disrupted this image; they have allowed for a number of groups each to claim to be the legitimate heads, the legitimate centers of thinking and creating, and hence to claim that they ought to be granted a status different from the one held by the others. In the past, cartographers could claim that it was they who constituted the locus of thought and creativity, and that those who provided equipment and data were merely adjuncts who provided labor (products themselves being the outgrowth of labor). But the development of geographic information systems has given these others an opportunity to claim that in various ways they are central, and that they ought thereby to be accorded rights and responsibilities previously reserved for others.

As a result, we are on the brink of a wholesale refiguration of the system of rights and responsibilities. Unfortunately, those in the academy who are involved in the use of the systems have failed to notice these developments because they have so often seen their own positions in terms of that firmly entrenched guiding image. What follows is an attempt to make clearer the nature of that image, the way in which it has been challenged by the development of geographic information systems, and the implications for the system of rights and responsibilities.[33]

What follows will be in three parts. First, I shall lay out what I take to be the usual way of thinking about rights and responsibilities within the context of the production of science. Although geographic information systems are widely used outside of the academy, they are so often used in scientific research that the model of rights and responsibilities used in science is especially appropriate. Here I shall argue that within the academy there is a strong presumption that we can separate work into that of the head and that of the body, and that we need to allocate rights and responsibilities on that basis. This process works within the academy, but it also works outside it; those in the academy typically represent the actors outside of the academy, in government or industry, as bodies to science's head.

Second, I shall point to the ways in which as a result of the development of geographic information systems, the academy has been faced with other actors who themselves are able to claim the status that academics and scientists have long claimed. Now those in industry can claim that it is they who are the locus of thought and creativity. Indeed, even within industries there are now disputes over which part of industry is the essential part, the locus of mentation, and which merely the laboring body.

Third, I shall show how the use of geographic information systems is tied to a recasting of understandings of rights and responsibilities within the discipline.

A geographic information system is a complicated thing. This may seem a truism, but in saying it I mean more than just that a geographic information system (GIS) has a great many parts. I mean, too, that it consists of a diverse set of elements, used and operated on in a diverse set of ways; a GIS is at the center of a wide range of practices. This turns out to be important because it is in the context of those practices that patterns of rights and responsibilities develop, and the diversity of practices leads to a diversity of rights and responsibilities.

In the last chapter I described some features of two main ways of thinking about the right to intellectual property, one often termed the "labor theory" and associated with John Locke, the other termed the "personality theory," or theory of moral right, and associated with Hegel. Here I shall look in more detail at the ways in which the distinctions made there are played out in the case of geographic information systems.

In the academy we are especially concerned with the rights and responsibilities that attend the practice of science and organized inquiry. The responsibilities are perhaps the more familiar of the two. We have a responsibility to be honest, to mean what we say. We have a responsibility to give credit when we use the work of others. And we have a responsibility to make our work available to others, to share it. At the same time, we have a series of rights, which are in a sense the flip side of the responsibilities. We have a right to expect others to be honest. We have a right to be given credit when others use our work. And we have a right to use the work of others without first asking as long as we give that credit. These are moral rights and responsibilities. Indeed, the system of science operates on the assumption that it is at its heart a moral system, that its practitioners can be counted on to engage in a set of practices that exemplify a set of values, of honesty, altruism, and communalism.[34]

At the heart of this system is the written work, the article or report or book, or perhaps more recently the E-mail message (although this raises difficult questions). Because the written work is a report of the findings of the scientist, when one scientist cites (or fails to cite) another, it is really the written work that is the focus of the action. Similarly, when we speak of a scientist being honest, we typically are speaking of the scientist's words only insofar as he or she is speaking as a scientist, and the written work is the place where a person can be held most accountable.[35]

This process is not, though, as simple as it might seem. Consider a "regular" academic paper. On the traditional image, I do the research, write it up, publish it, and get the credit. I may acknowledge the assistance of other individuals and organizations, but on the whole both the rights and responsibilities are mine. Now, this does seem simple. Yet as we have seen, there are really two ways in which this characterization of the attribution of rights and responsibilities appears on closer analysis to diverge from what is really happening. First, in almost any conceivable set of circumstances there are a number of other people who have been involved in my work. There may be secretaries, lab technicians, research assistants, and editors; at a somewhat greater remove there are typesetters and administrators, copy machine technicians, the postal service, and tax collectors. One could, of course, follow the ripples on out across the society. To the extent that they take part in that work, these individuals, groups, and organizations are to varying degrees responsible for the work of the scientist. And yet they are seldom given credit, even more seldom given real credit, and almost never asked to shoulder the responsibility for inadequacies in the research carried out with their help.

There is a second way in which the usual characterization of rights and responsibilities in science diverges from what "really happens." And that is that few of us actually own the works for which we claim both rights and responsibilities. By and large, when we come to an agreement with a journal or book publisher, we agree to sign our rights over to the publisher. This means that the publisher can decide when and how to publish it, whether to reissue it, and even whether we ourselves can make copies of it.

So in two important ways the usual image of science diverges from the practice of science. In fact, one of the interesting things about science—and one of the things that needs to be explained—is the way in which scientists, and the public, have blithely gone along with this usual image, accepting it as though it is accurate and complete when it patently is not.

I have noted that both views of rights and responsibilities are widely held; in fact, it is not inaccurate to say that both are seen and invoked as common sense. This certainly is because a version of this view is at work in everyday conceptions not only of the work of scientists but also of work more generally. As theorists from Karl Marx to Frederick Taylor to Harry Braverman have pointed out, in the division of labor as it operates in our society, there is a distinction between management and labor, between thinking and doing.[36] Indeed, in the factory as in the laboratory, there is a similar mapping of the head and the body, and a similar ascription of one form of rights and responsibilities to those thought to think, of another to those thought only to work. There, too, the portion of work associated with thought and creativity is seen as essential and worthy of permanent credit (and blame). The part associated with labor is seen as contingent and replaceable, and rights and responsibilities are shed at the door.

Yet the example of the factory brings up a crucial issue. For in industry, perhaps more obviously than in the academy, not all thinking counts as thinking. The person writing lines of computer code is surely engaged in thinking; yet this work, and the work of the digitizer, is viewed as a species of labor and grouped with that of the tool and die operator.[37] The difference is this: this sort of intellectual work is seen as routine, habitual, interchangeable. It is not merely thought that counts, but creative thought, thought that is somehow beyond routine. And in an important sense the evidence that this breed of thought is taking place is found in the status of the person involved; far from democratizing the workplace the computer has simply replicated old patterns of inequality.[38]

GEOGRAPHIC INFORMATION SYSTEMS AND THE DISINTEGRATION OF AN IMAGE

This image of the head and the body, of creativity and labor as alternative ways of gaining rights and accruing responsibilities, pervades the everyday practice of science. Many who are enmeshed in it, of course, see it as a natural system. And this view is supported by the fact that it appears to meet the interests of a number of groups of people. For example, scientists like the image because in allocating to them the moral right it is a means of justifying a permanent set of rights, and thus a permanent source of status. And many who are involved in more mundane aspects of science have tended to like it because it provides monetary benefits but a strongly attenuated set of responsibilities; many people, after all, wish to leave their jobs at work when they leave for the evening. Granted, there have been some who have not liked it; it has, after all, been a system whereby certain people, like graduate students and research assistants, and especially women and members of minority groups, have been denied credit. But for graduate students, who lead an ambiguous existence, there has always been the promise of more, and women and minorities have had little choice.

But as strong as this image is, when we turn to geographic information systems, matters seem confusing; it is hard to know how to apply that image. One might argue that the image does apply, and in an obvious way, where a scientist or planner is using a system designed to be a black box, a self-enclosed entity that operates smoothly and without intervention. There the role of the user looks very much as though it involves all manner of creativity, as she sets system options through a set of menus, and then creates end products (whether in the form of analyses, maps or other representations, or systems of directions).

Still, if this sort of work is clearly not "just labor," neither does it quite fit our usual notions of mentation and creativity. In fact, it might be argued that here we are moving into another realm of imagery altogether, where the person using the system can be seen as a driver, or perhaps better, as the helmsman of a ship, there to guide it to a destination determined by the client. In this case, with respect to practices, the user of a GIS seems to share a great deal with the technician in the physician's office; both use efficient tools that have routinized what otherwise might require difficult and time-consuming labor. Although both are involved in a labor process, it is a highly skilled process. Indeed, this imagery leads us into a completely different way of thinking about rights and responsibilities, to the way used in studies of professional ethics.[39]

Nonetheless, I would argue that the image of the helmsman does not really capture the ways in which geographic information systems are used in research. At the same time, I would argue that the traditional image of science seems no longer to work, that we can no longer easily distinguish a single creative head and a laboring body. Rather, there now is a complex, hydra-headed beast, with the various heads arguing with one another about which is the "real" one. In the remainder of this section I shall spell out what I mean by this and what the consequences are for those involved in the practice of developing and using geographic information systems.

The best way to show what I mean is by looking at the most obvious example. Imagine that you are discussing a person's work, and that that person does work in cultural geography. If you ask, "What kind of word-processing software do you use," you are extremely unlikely to believe that from the answer, "I use WordPerfect 5.1," you can draw conclusions about what sort of cultural geographer that person is. Nor is that person likely to fear that at some point in the future the WordPerfect Corporation will come back and ask for royalties for articles and books written using its software; that is, it is extremely unlikely that anyone would assert that the use of WordPerfect 5.1 was essential to the carrying out of the project.

By contrast, we actually do feel justified when we hear "I use ArcInfo" or "I use Atlas*GIS" in drawing some conclusions, at least conclusions about the limitations of the sorts of work that can be done. And as a result, here we may have the feeling that the software itself has become an integral part of the process; the situation may seem very much like the one that we find in filmmaking, where the credits list the name of the camera, the film, and the lenses. Indeed, we might very well argue that unlike word-processing software, GIS software is absolutely essential to what gets done.

If software can be seen as essential in the case of the smallest and simplest systems, where the output is in the form of simple two-dimensional maps or data analysis, it becomes more essential as the size of the software system increases. And it is not simply that the creator of software has a right that the work that she has created not be plagiarized, that the scientist not create "knockoffs," or copy the "look and feel" of software, or manipulate the code in order to make the product better suit his own needs.[40] There would be little disagreement that in such a case the creator's rights were being violated—although some might want to claim that in this case the rights are primarily monetary. Rather, the issue of rights arises in every case in which the scientific work being performed is such that only software of a great size can perform the analysis (or, perhaps, can perform it within the given time constraints). There, I would argue, the claim that the software is essential to the project and that therefore the developer of the software ought to be given equal right appears quite credible.

If this is the case with systems of great size, it is perhaps more true of systems of great complexity. For example, one can make the same argument in the case of systems based on artificial intelligence; there the rules that comprise the system are so complex that no single person could conceivably run through even a single iteration.[41] Under those circumstances, it does not seem at all unreasonable to assert that the work could not have been done without the software, and then further to assert that one needs somehow to give real credit to the producer of that software. This is especially the case with artificial intelligence systems in which software developers have argued that the "personality" of the system ought to be patentable, but I would suggest that it applies to complex systems more generally.[42]

One rejoinder might nonetheless be that in principle one could even in these cases of great size or complexity produce the results by hand. One might argue here that the case of software is very much like that of laboratory apparatus, that if a complex and proprietary piece does the job better, a simpler and more generic piece might at least do the job with a little (or even a lot) more time and effort. So it might be argued that the giving of credit is in such cases perhaps not inappropriate but at least inessential.

But even if we grant that, it seems difficult to imagine being able to make this same argument in the case of certain more recent and proposed applications of geographic information systems, where the output may be in the form of images or even videos or sound. There the possibility of operating in real time imposes constraints that cannot easily be avoided.

These complex software systems suggest that the same argument might be raised with respect to hardware too. It would seem more than a little odd if the Koh-I-Noor company were suddenly to begin arguing that it ought to receive credit for all maps made with its pens. Here, after all, it could be replied that as in the case of word-processing software, the use of this or that technical pen is purely contingent. But when we turn to systems that are extremely data intensive and require inordinate amounts of rapid access storage, or which require highly specialized technologies for the representation of results—again, for example, in the case of audio and video and especially of virtual reality—we may find that the hardware itself has become the centerpiece of the system, that while certain features of what is being done can be imagined to have been different, we cannot imagine the system without its very specific hardware. This very argument was recently advanced when a computer manufacturer attempted to submit a bid as a general contractor for a government aircraft; the manufacturer claimed that the computer was the central part of the aircraft, and that the airframe existed simply to hold it up.

Finally, if we turn to the data that are being analyzed, we may find a similar situation. If we are using data provided by others, there may be cases—even if those data are public and might very well have been collected by us—where some combination of the means of ordering and the means of selection would appear to give to the provider of those data a special right. Imagine, for example, a very detailed survey of an area carried out using a satellite imagery or a global positioning system. Indeed, it is this very view of the centrality of data that has in the recent past been invoked in a wide range of attempts to rewrite intellectual property regulations in Europe.[43]

So in the three central cases, of software, hardware, and data, there are good reasons for arguing that the providers of those elements of a geographic information system have contributed parts that are absolutely essential to the operation of the system. And it makes sense to argue that as a consequence they have acquired sets of rights that are on par with those traditionally ascribed only to the scientist.

If we turn to the other side of the matter, to the ascription of responsibility, we find very much the same situation. This is perhaps most obvious in the case of data. If in an article I include a reference to another work, those who read my work have every reason to expect that my reference is accurate and germane. If it turns out that I have referred to a work that does not exist, the reader has every right to claim that I have been irresponsible. Similarly, where survey research is carried out by a team, it is usually assumed that the head of the project has ultimate responsibility for the project. It is assumed, that is, that the project head has taken due care in determining that the people doing the work are qualified and responsible.

But when we turn to extremely large-scale databases, the situation is surely different. If I purchase a set of data from someone, I am in a position to claim that those data come with a claim of accuracy, and that any errors or omissions within those data are not my responsibility. Furthermore, I am unlikely to believe in a case where a large data set that I have bought or rented includes individual items that were themselves obtained through a process of violation of the rights of certain individuals, that I ought to be held liable for that violation. Legally, an attribution of responsibility here would require that I had been able to foresee that the data were improperly obtained.[44]

And even when the database has not been purchased from a commercial vendor or procured from a government organization, when that set of data reaches a certain size, we feel justified in claiming that just by virtue of investing that much labor in the database, the author has implicitly made a claim for its value.

This issue of the relation between sizes and responsibility is well known in the case of data; the development of the sorts of extremely complex works of software used within geographic information systems now raises the same issues. Here, though, we need to distinguish computer software—and even complex software—used for simple tasks, like word processing and spreadsheets, from that used in the more complex environment of a geographic information system. For in the case of a word-processing program, and even an extraordinarily complex one, it is easy to monitor the relationship between what is put in, in the form of keystrokes, and what comes out, in the form of formatted text. But as this relationship between input and output becomes more complex, as in the case of GIS software, it becomes much less simple to engage in that sort of monitoring. And this is all the more true because even if the user of a geographic information system is familiar with the general outlines of the software being used, it is far from likely that that user has the same familiarity with the ways in which all of the analytical operations that are stored in the software work.

If this is true of traditional statistical operations, it is all the more true when we turn to the more advanced systems being developed today, and especially to those systems that incorporate artificial intelligence. Indeed, where the aim of the system is to incorporate into the software the ability to learn, one might want to say that the ability of an individual to grasp the appropriateness of the decisions prescribed by the computer is evidence of the failure of the software.[45] Here too, then, when it comes to the issue of responsibility, it becomes difficult for someone who uses a geographic information system to take responsibility for the way in which the software works in the same way that that person could take responsibility in other situations.[46]

Finally, much the same argument holds in the case of hardware, most clearly in the case where hardware has been designed to mimic the operations of software, which has, of course, been designed to mimic the operations of the human mind. Here the development of computers has occluded the distinction between choices as made by individuals and described by them in natural language, choices described in a formal language (like symbolic logic), and choices made, metaphorically, by electronic circuitry.

In effect, then, in the matter of responsibilities as in the matter of rights, geographic information systems involve a refiguration. The locus of responsibility has shifted from the user to the producer of software, hardware, and data. Whereas before it made sense to think in terms of an image of the scientist as a thinking/creating head, leading the rest of the team, it now makes more sense to see the "head" as variously located, and even to see science as a multiheaded object, with other heads outside the academy, located in industry and government.

Through the popular press most of us are familiar with the consequences of these changes. The sordid side has recently been widely reported in stories about the home of bureaucratized technologized science, medical research, which has in the past few years seen a series of scandals.[47] The most famous of these concerned alleged fraud on the part of the main author of a paper that had multiple authors, some of whom had apparently never even seen the offending piece but nevertheless were given equal billing.

In one sense, of course, this example supports my contention about changing rights and responsibilities because in medical research it is quite common to give authorship credit not only to active participants in research but also to a range of others, including the person responsible for running the laboratory where the research was carried out.[48] Thus with respect to the extension of the rights of personal authorship, the situation in medical research is very much like that in geographic information systems. And medical research is similar in another way as well. In medicine and biology articles frequently read like technical manuals for repairing an automobile; they include explicit directions, including the types of equipment used. Here, too, rights are being shared.

Yet in the case both of medical research and geographic information systems, the following claim might be made: both typically involve the asking of questions, the marshaling of resources, the drawing of conclusions, and the attempt through the process to create representations of what is real. And in both cases, and notwithstanding the use of and even deference to technology, there remains the view that the aim of research is the discovery of a permanent truth, which can be characterized eponymously, as Smith's principle or Wasserman's discovery. And in this act of eponymy the technology disappears, as rights and responsibilities migrate to the person who is given credit, while those of others come to be devalued, seen merely as contingently related labor. Hence, it might be countered that the apparent importance of other actors, other "heads," is merely a temporary feature of the research process. In the long run we are always able to advert to the familiar image of the head and the body, to locate the permanent moral rights and the attendant responsibilities where they belong, with the scientist.

This argument, though, rests on another, and clichéd, image of the nature of the practice of science. On that view the goal of science is the creation of theory, and we can leave aside issues of these socially and economically situated technologies because within science they lose importance in the face of major theoretical discoveries. But this view needs to be rejected. Here the cynical geographer might say that it ought to be rejected simply because we have no theoretical discoveries to study. But to see a more fundamental reason for the rejection of this premise, we need only look to recent events in biology and medicine, where artificially created forms of life have been patented. This and other developments in the attribution of rights in science can only lead to the conclusion that the areas within which it is possible to think of oneself as doing "pure" research are becoming increasingly limited. When the right to a discovery or the right to reveal data is held by a corporation, it is difficult to see how the scientist can claim real independence. And here geography, in fact, provides a more extreme case because it can and must rely so strongly on large-scale sets of data and on data provided by government (and now industry). Here the image of the head and the body remains a way of obscuring the real structure of geographic practice.

And yet the development of new sets of relationships among the university, government, and industry has been so substantial, and the boundaries between them have become so porous, that it is now difficult to imagine a clear guiding image. Indeed, it is this difficulty that makes our ruminations about rights and responsibilities so taxing. It is as though all of the landmarks have been removed.

Conclusion

And so, the development of citation indexes, large-scale funding, and computer mapping and geographic information systems has in each case been associated with what can only be described as a refigured role for the written work and a refigured place of the written work within the academic workplace. In one sense this is not at all a surprising conclusion. After all, much of what I have cited as evidence is there in plain view, there to be seen. But it remains partial, disconnected, unless we are able to see the written work within the context of a series of places. Indeed, it is only within those places that the written work is a work. It is not simply an abstraction, but incorrect, to imagine that one can see the work as anything other than an object that fundamentally exists only within such places.

But once we begin to take seriously the need to see the place within which the written work is used, we begin at the same time to see those works as existing within what can only be characterized as places that are both unstable and rapidly changing. And we cannot help but see that the written work is itself changing as those places change. In an important sense what we are seeing here is the destruction of the written work as an object and its reconstruction as an armature around which operates a system of surveillance, a system of technologically sophisticated work processes, and a set of competing rights and responsibilities. And as such, the work—as we shall see in the next chapter—loses much of its freedom to represent.

CHAPTER SEVEN

Finding the Space in the Text and the Text in Space

In recent years a new group of geographers have taken up the question of the nature of space. Taking as their target the "dead" space of Newton and this time adverting to the work of Henri Lefebvre, they have argued that by looking at the process of the "production of space" we can once and for all get clear both about the nature of space and about the sources of past theoretical failures. If Lefebvre's supporters are right that he was right, there is much reason to celebrate; what is left is a process of mopping up. But it should be clear from what I have said that I believe that they are not right. Indeed, one might see this book as an attempt to understand just *why* they are not right. As it turns out, the problem is *not* that Lefebvre has done a bad job at the task that he set himself. Rather it is for another reason; it is because the task itself is an impossible one.

The task would, of course, be difficult enough just to the extent that it assumes it possible to create a theory of space that both explains space and explains itself. But this point about reflexivity, one often made by critics of grand theory, is not the end of it. For more fundamentally, the project must fail because in the process of "doing the theory" of the production of space those involved make use of and appeal implicitly to a variety of theories or conceptions of space. Indeed, this occurs inevitably simply because of the social, political, and economic situations within which this project of theorizing takes place. More specifically, because *The Production of Space* is a written work, Lefebvre has involved himself in a series of commitments to a set of conceptions of space firmly entrenched within Western thinking. In making those commitments he has undercut the possibility of developing a thoroughly critical view, because through them he has honored those very views that he has sought to criticize.

Now, it may appear that in saying this I am singling out Lefebvre and his followers. And to be honest, to the extent that Lefebvre's followers have pictured him both as the first person to have seriously engaged the issue of space and as the creator of a well that contains all that is important to be said about the issue—while he remained silent about the written work—they set themselves up for criticism and critique. But although my argument implicitly addresses points made by Lefebvre and his followers, this ought not to obscure a more important point, that insofar as it is presented in a written work, a theory about space—or about anything else—in *every case* entails the involvement in a set of practices and institutions, and different practices and institutions are structured in terms of different conceptions of space. We can always find the space in the text.

The chapter has three parts. First, I shall briefly lay out the lineaments of Lefebvre's work. Second, I shall note the ways in which his project involves appeals to a wide range of conceptions of space. It does so in its conceptualization of author, reader, and text, and also in the way in which the written work itself fits into a larger set of institutions. Finally, I shall then show how these many conceptions of space can be seen as fitting within the much smaller set of basic conceptions described in chapter 4.

Looking at Lefebvre

Lefebvre's work appears in every way shot through with the recognition of the importance of space and with attempts to address its nature. And it is not as though his is a brief attempt; for four hundred pages he takes one tack, then another. And yet, in the end I think that one has to feel a sense of disappointment in his work, a sense that he has not quite succeeded.

There is a tendency to assume that a work of geographical theory fails for what turns out to be a simple reason—the theory does not work. It may be internally inconsistent, or may fail to "fit" the facts, or may be irrelevant or banal. But Lefebvre's difficulty, I would argue, cannot be adequately apprehended if we treat his work in this way, if we look simply to its representational success or failure. Rather, we need to look at the work more broadly, to see the extent to which its conceptualization of space is both inconsistent and constricted.

Lefebvre's account of space is complex, approaching the issue first from one angle, then from another. From one perspective he argues that we can see space in terms of three "fields" or levels, the mental, the physical, and the social. From this point of view one needs to see the understanding of space as deeply constrained by the privileging of mental accounts given by philosophers and scientists, which see questions about space as having been solved, and see space as a kind of abstract container. Here mental space is seen as abstract, imaginary, and as something to be contrasted with real space, with the social and the physical.

At the same time, people interact with space in three ways, and this produces three "types" of interactions. People engage in spatial practices, and there the mode of relationship is perception. They make representations of space through conception. And they carry out their lives in day-to-day contact with spatial representations; their relationship to these is seen simply in the process of living.

There is a history of human relationships with space, in which we can see broad changes. At one time people lived in relationship to absolute space, to "fragments of nature located at sites which were chosen for their intrinsic qualities."[1] But—and in this historical conceptualization Marx comes to the fore—

> the forces of history smashed naturalness forever and upon its ruins established the space of accumulation. . . . It was during this time that productive activity . . . fell prey to abstraction, whence abstract labor—and abstract space. (49)

But like the capitalism that engendered it, abstract space has built within it contradictions, and ultimately it will collapse. In its place will rise a new "differential space":

> [I]nasmuch as abstract space tends toward homogeneity, towards the elimination of existing differences or peculiarities, a new space cannot be born (produced) unless it accentuates differences. It will also restore unity to what abstract space breaks up. . . . (52)

Finally, it is important to see that forms of space constitute codes, "a language common to practice and theory, as also to inhabitants, architects, and scientists" (64). These systems of implicit rules or practices are multiform; citing Barthes, Lefebvre speaks of five: knowledge, function, symbol, emotion, and critique (161). Each of these aspects of the question is relevant,

Lefebvre argues, to the long-term development of space, to the space in which people live, to their beliefs about space, and to the theories entertained by scientists and philosophers. Once we begin to look at space in this way, against the background of Marx's understanding of the development of class, capital, and the forces and relations of production, we can begin to see developing patterns, to create a clearer understanding, even a science, of space.

Now, Lefebvre has offered up a grand theory here, one in a long tradition of grand theories. As I have characterized it, it seems extraordinarily broad. And for that reason it may seem a bit odd to suggest that one of its problems is that it is too narrow. But that, I shall suggest, is true. Moreover, one may wonder what it might mean to say that it is inconsistent, because on the basis of what I have said there is little reason to make that claim. But that, too, is what I shall suggest.

Text as Representation

THE AUTHOR, THE READER, AND THE TEXT

Authorial Space The creation, in Lefebvre, of a set of spaces begins with his creation of an authorial space, or to be more accurate, of three different and conflicting authorial spaces. He begins the volume as follows: "Not so many years ago, the word 'space' had a strictly geometrical meaning: the idea it evoked was simply that of an empty area." Then, "not that the long development of the concept of space had been forgotten . . ." (1). And on to the very end, where he says, "Space assumes a regulatory role when and to the extent that contradictions—including the contradictions of space itself—are resolved" (420).

This is the first of the three spaces; from beginning to end Lefebvre plays the part of the neutral author. It is not that "I thought that space was such and such," or "It seems to me that people believed such and such," or even "I have found it important to understand that such and such." Rather, the author sets out a set of facts, critiques, and theories as though they come from an impersonal observer, one standing outside of that which is being described.

It is not, though, exactly that the author is standing utterly outside of what is being characterized. One is met first with one scheme, then with another; if each is claimed to divide the conceptual world with no residue, together these schemes, coming one after another, suggest—at once—

that all of the ground has been covered and that faced with a difficulty, the author would pull another scheme from his hat. And because in an important sense his way of characterizing the world leaves nothing "outside," what we are really seeing is a view from above, a view in which the author expresses his authorship from a position not exterior but superior.

This view of space, the space of the observer, sits uneasily alongside a second sort of authorial space, one that is itself ambivalent. Here the author takes and holds a position against others, who are doing the same. Now, on the one hand we can see this as a matter of the establishment of a position in conceptual space. On this view there are a variety of positions to be had, or held, and some are closer and some less close one from another. For example, Louis Althusser and Manuel Castells are closer to one another than is Castells to a member of the Chicago school of sociologists. On this view one is located in conceptual space in a number of ways, as a result of training and education, politics, and so on. And it is possible to create new positions within this space, even to alter the space itself. Once one takes up a position, though, one holds that position, very much as a military battalion holds a position against outside forces.

Now this view, where academics is seen as a series of battles over positions, fits hand in glove with another. Here the author refers to those others as though they were merely standing offstage. The Marxists, the positivists, those who see space merely as a neutral container, all are standing in the wings, waiting to respond. Christian Norberg-Schulz postulates, Victor Hugo envisions, all are there, waiting to respond, or waiting for Lefebvre to construct their responses for them.

There is a third sort of authorial space in Lefebvre, one on the face of it very different from the other two. This is the space created by the author, as person. In fact, the author writes the work as though it has no author, or at least no human author. The "author" has no age, no gender, no race. The author is a citizen of no place, a resident of no place. The author is merely a source of ideas, a point source writing from an impossibly small location. It would be too much to say that the author writes from a point created by the intersection of lines in Euclidean space: the author writes from nowhere.[2]

And so, even as he ends, as he reflects on the entire project, we are left only with the following:

> The creation (or production) of a planet-wide space as the social foundation of a transformed everyday life open to myriad possibilities . . . is the same dawn as glimpsed by the great utopians . . . by Fourier, Marx and

> Engels, whose dreams and imaginings are as stimulating to theoretical thought as their concepts.
>
> ... We are concerned with ... a direction that may be conceived, and a directly lived movement progressing towards the horizon. (422–23)

Indeed, here we have all three views in a single paragraph, the glimpsing, the theoretical thought, and the anonymous "we." There are alternatives to each, but in Lefebvre the existence of those alternatives is obscured.

The Space of the Text If we turn to the text itself—leaving aside for the moment a set of semantic, social, and economic issues—we find another set of views of space. First, and clearly related to the issue of authorial presence, within the text Lefebvre claims authority (we shall see, later, that this is not the whole story) not because of social status or personal history, not because of gender or race, but—since nothing else is left—because of the power of the argument itself; that is, his attempts to convince the reader are based on a generic reader (I shall return to this issue), and a reader who can be convinced by argument and assertion rather than by force or deceit. And so, he says,

> What is commonly referred to as the "political question" needs to be broken down, for like space itself it gives rise to a number of sub-questions, a number of differing themes or problems; there is the question of the *political sphere* in a general sense, and of its function in social practice....
>
> That relationship, which has always been a real one, is becoming tighter: the spatial role of the state, whether in the past or in the present, is more patent. (377–78)

This all appears relatively straightforward: People refer to something as the political question. But that question, by itself, raises other questions. There is a political question concerning the relationship of space to the state. The nature of that relationship is becoming clearer.

But note, here, that we are not told anything about the author. Nor are we told about the "others" who have named the political question. Nor are we told why or how one question can engender another. And nor are we told why we should believe that there are changes in the relationship of space to the state. All of these ideas are asserted with the same air of flat, passive neutrality. It is this air that suggests that the author is approaching the world with the same sort of neutrality.

The volume itself supports this air of neutrality. Far from rambling and incoherent, it is introduced by its author's "plan," a seventy-page map of the volume. And the contents themselves give us further evidence that this is a work of manifest order; chapters move "from absolute space to abstract space" and "from the contradictions of space to differential space." Finally, when we turn to page 1, we find that we are immediately placed right in the heart of the matter: "Not so many years ago, the word 'space' had a strictly geometrical meaning..." (1). So the work itself *looks* like a serious, scholarly, straightforward attempt to get at the nature of space.

Further, one finds the usual accouterments of scholarship. There are footnotes. Some cite, but some explain; Lefebvre tells us that "so long as the focus of the discussion is on architecture, the best discussion is still E. E. Viollet-le-Duc" (119 n. 8), and in that way lets us know that he knows the literature. We are told that "according to a well-known formulation of Marx's, knowledge (connaissance) becomes a productive force immediately, and no longer through any mediation, as soon as the capitalist mode of production takes over" (44). And through the reference to this passage, "Karl Marx, Grundrisse, tr. Martin Nicolaus (Harmondsworth, Middx: Penguin, 1973)," it is made clear that he has read and grasped the entire work—as should we.

Now, if we consider the ways in which Lefebvre refers to the works of others, for example, "Chomsky unhesitatingly postulates" (5) and "Céline uses everyday language to great effect to evoke the space of Paris" (14), something else emerges. Writers whose work is twenty, thirty, even forty years old, and who are long dead, are referred to in the present tense. It is only when we move from the near present back to the 1940s that the way of referring to writers begins to waver, so that ultimately references to Leibniz, Descartes, and Marx are in the past tense.

This use of tenses reflects an implicit, and scientific, notion of the relationship between author and ideas, where in a sense the author is drawing on a current "stock" of ideas, there to be chosen from, and where referring to an author in the past tense does not simply place the author in the past but rather signals that that person's ideas are not part of the current stock. To use those in everyday work is to be distinctly antiquarian, and this will not do, because the goal, at least implicitly, is to develop an account of space based upon and incorporating contemporary theories.

Finally, the work ends with an afterword by David Harvey. Harvey calls Lefebvre's work "magisterial" (425), offers a biography, and concludes with a list of sixty-six books written by the author.

And so, from start to finish the reader is made to develop an expectation. The work is represented as a presentation of a set of ideas. The ideas are the result of a serious attempt by the author to see what space is really like. The volume will be not a polemic, not a diatribe, not a work of fiction, but a serious engagement with issues related to space.

As he creates this expectation, Lefebvre situates the text in a neutral conceptual space, and he does this in several ways. He represents it as a neutral vehicle for the ideas of a neutral author. It is a vehicle that claims the authority of neutrality by associating itself with a series of traditions of format and style. Although in the citations that he uses Lefebvre associates himself with a certain group of thinkers, with Marx, for example, the ways in which he uses citations, and his traditions of format and style more generally, place the work within a tradition that views thoughts (and I use the ocular metaphor intentionally) as elements that can be located in a timeless conceptual space.

The Space of the Reader Finally, Lefebvre creates a space within which the reader exists. As with authorial and textual space, he does this in several ways. First, he addresses the reader as though that person (he?) is an individual and a silent reader. This is a text that in a fundamental sense is written for a reader who is outside of time and space, where the reader can imagine the text as a medium providing direct access to the author's thoughts, which are themselves outside of time. In this sense the text is a kind of mirror, capturing Lefebvre's thoughts, catching them and reflecting them into the reader's mind.

Lefebvre writes as though the entire text exists as a whole in a textual present. There are several ways to put this, but perhaps put most simply, the text, from beginning to end, is given to the reader all at once. It is a whole, set down in front of the reader with the implicit message, "This is what Lefebvre thinks about space."

Perhaps the easiest way to think about what this kind of text means is to consider an alternative, an intellectual diary. Imagine that you decide to give some thought to the nature of space, and the first entry begins: "Jan. 1: Decided to think about space." The reader of your daily entries will almost certainly find a kind of grab bag. There will be themes carried for days, even months or years, themes started and soon abandoned, themes clung to long after they have ceased to bear fruit. The structure of this text might very well be seen as something like a journey, with a beginning and goal, a middle, and an end.

Now it may seem that this kind of text is not exactly what we find in academic prose, but in fact it is just what we often find. In geography a classic of this sort is Harvey's *Social Justice and the City*, with its move from an early liberal to a progressively more radical position. We find this sort of text most often in the form of collections, some of whose authors are quite explicit in using the genre as a means, retrospectively, of rethinking their intellectual paths. But Lefebvre has not done this; he has offered a text that claims to be written outside of time, and thereby asks to be read outside of time. And to be read outside of time is, too, to be read outside of space and place. It is to be read by a reader who is a naught.

But the reader is a naught in another sense. Writing within a scientific tradition that sees rhetoric as a bad thing, Lefebvre operates as though the job of the author is to demonstrate—but not to convince. The conceptual and theoretical distinctions are supported by assertions of fact, and the argument follows an order; in these and other ways the author writes within the conventions of science and history and develops a rhetoric that claims to be neutral. The reader, that is, is not explicitly conceived of as an individual who has interests to which appeal should be made; rather, the reader is conceived of as a person for whom rational demonstration is enough.

And so, in both cases an imagined reader—ageless, genderless, raceless, placeless—is constructed. Just as the author is in one sense represented as existing outside of time and space, so too is the reader, one who reads from a zero point. And the reading takes place beyond the world of interests, in one where there is no convincing but only demonstration.

Alongside this, though, there are other notions of the place of the reader. For example, the argument proceeds less from the overwhelming use of evidence than from the laying and relaying of conceptual schemes. There are only 165 footnotes in a 423-page manuscript, and with respect to the issue of space itself, the citations are especially meager; the text rolls along for over 400 pages with unsubstantiated and undocumented assertions. In this way the rhetorical strategy of the volume is less to demonstrate than to force the reader to submit. It is as though every objection can be answered by an assertion to be found somewhere in the text, every conceptual problem can be resolved through the appeal to another conceptual scheme. The reader is placed in a position where any objection can simply be incorporated into the text. The reader becomes an unwitting collaborator. And here, the reader becomes, too, an unwitting part of a community, a group who in grasping Lefebvre or offering to be used by him set themselves aside from the rest, set themselves into a special, intimate space.[3]

In fact, far from being one that can be read outside of time and space, Lefebvre's volume is written for, and needs to be read within, a very *special* time and place. This is not a book to be read aloud, not one to be read in groups. It is not a text to be read at the barricades, not one to be kept alongside the die grinder on the workbench. And it is surely not one to be kept next to the crib. Rather, it is a book that demands a time and place devoted exclusively to its reading, and this is a time and place exclusive to academics and to others of privilege. By now, of course, it is a commonplace that one of the methods that academics and scientists, and professionals more generally, have used to establish their authority—and a method that was a central element of modernist culture—was to claim for themselves an exclusive time and place, outside of time and space. To be outside of time and space is, after all, to be beyond the control of intellectually deforming interests.[4]

But it is more typical today to argue that there can be no neutral reader, no reader outside of time and space. Whatever their relativist excesses, in the last fifteen years social constructivists have rather completely destroyed the myth of the interest-free scientist.[5] In its place they have offered a view of scientific production in which the play of human interests reaches all the way down, all the way to those pursuits, logic and mathematics and physics, that once appeared to be incorruptible bastions of neutrality. The focus of this work has largely been on production, including the production of texts.[6] But it has implicitly, and I stress only implicitly, made the same claim for the reader, who now appears to read from a position in time and space. If Lefebvre grants this in his text, he denies it in the way that he writes that text.

When we pick up a book like Lefebvre's, then, we are in a number of ways told about it, about its author, and about how to read it. Indeed, by the time we get to the *point* of reading Lefebvre we have long since learned that writing, and more so reading, are natural acts. To write is to put onto paper your thoughts; to read is to pick them off of the paper and place them in your mind. Here the book, the physical repository of language, and hence of thoughts, is a kind of intermediary, which in all but the worst of cases acts transparently.

But as we have seen, on closer examination these relationships are not as clean and clear as we have been taught to believe. In the case of Lefebvre, we find the author representing himself in three very different ways, as an observer outside of space, as the holder of positions in a kind of conceptual space, and as an anonymous individual within space.

The text itself presents itself as unproblematic, as just another book. It has all of the accouterments of such a book, and because of that is able to maintain an aura of neutrality, as though it is simply another set of representations in conceptual space. At the same time, and despite his avowedly historical interests, in his ways of writing, and especially in his use of tenses,

Lefebvre creates a bifurcated world, with his position and those of his opponents situated in a textual now, and the past marginalized.

Finally, we find a complex set of readerships. In one sense Lefebvre simply works within the conventions of the text; the reader is viewed as a neutral receiver of the author's ideas, transmitted through the medium of the text. Similarly, his traditionalist abjuring of rhetoric, his appeals to demonstration and proof and his refusal to couch his arguments in terms of the explicit interests of the reader, constructs a reader who is simply generic, anonymous and featureless. But beyond this rhetoric of no rhetoric there lies a different approach. He constructs the text in a way that piles elaboration upon elaboration, scheme upon scheme. It is a way that redefines the reader as collaborator, that defines an author-reader community separate, in a different space, from the other. And he constructs a text that notwithstanding its claims is written for a special time and place and for the reader able to fit within that time and place.

So in fact, when we consider Lefebvre's text simply in terms of its relationships to the author and the reader, we find a complicated story. Text, reader, and author are represented, explicitly or implicitly, as occupying a complex and contradictory set of spaces, hierarchical and nonhierarchical, temporal and atemporal, individual and communal, interested and neutral.

But this is not the whole story, for the text is itself a part of three other nexuses. It exists as an element of a nexus of ideas-language-text-world. It is an element of a nexus that combines text and social structure. And it is an element of a nexus that brings together the text as commodity and property. As we shall see, the development of an understanding of these three additional nexuses shows Lefebvre's work to be an element of an even more complex and contradictory set of spaces.

THE WORK, THE IDEA, AND THE WORLD

I suggested earlier that a simple way of thinking about the relationships among world, idea, language, and text—and a way that is commonplace in the modern age—is this: I have ideas and concepts in my mind, the idea of space or time, or the concept of a table. There are certain sounds, called words, that I arbitrarily associate with those ideas and concepts; because others who speak the same language use the same sounds, I can let them know what I am thinking by uttering those sounds. When they hear them, the sounds will conjure up in their minds the ideas that were in my mind. Of course, we have "names" not just for objects and theoretical entities but also for qualities, actions, and relationships, but this makes no particular difference. Nor does the fact that some words, like "he," refer by pointing to their objects rather than naming them.

On this view, written signs are very much on par with spoken signs. The difference is simply that in the case of written signs, I see rather than hear. In both cases, of inscribed squiggles and noises, we can refer to the entire body of signs used by a group of people as a language; a language is a set of signs. And this is to say that it is a means of communication, a means of getting one person to see what is (or was) in the mind of another.

I further noted that some authors saw such a view of language as getting in the way of understanding the workings of language and the process of representation; Richard Rorty has noted that in the last forty years a series of alternatives have been developed. Based, variously, on Wittgenstein, W. V. O. Quine, Gilbert Ryle, and the American pragmatists, more recent students of representation, language, and epistemology—including Rorty himself—have attempted to get beyond the spatial metaphors that defined the mind-idea-language-world nexus.[7]

So the questions that arise with respect to Lefebvre are these: given that he is interested in understanding the nature of space, does his understanding of language incorporate a critical perspective on space? Has he moved, like Rorty, away from certain spatial models of language? Is his view like Rorty's? And given the critiques that have been leveled at Rorty, has Lefebvre developed an alternative way of thinking about language that escapes some of Rorty's problems? Or has he retained, perhaps in a more subtle form, an unexamined commitment to a spatial view of the relationships among language, ideas, the mind, and the world, which colors the remainder of his analyses?

On the face of it, Lefebvre has responded to and rejected some of the views that we find rejected in Rorty. In an extended analysis of language he contrasts two approaches:

> At present, in France and elsewhere, there are two philosophies or theories of language.... According to the first view, no sign can exist in isolation. The links between signs and their articulations are of major importance, for it is only through such concatenation that signs can have meaning, can signify. The sign thus becomes the focal point of a system of knowledge.... (132)

This privileging of language, though, is dangerous; once one sees signs as the central issue in knowledge, the search for signs finds them everywhere:

> This search is assumed to begin with linguistic signs and then to extend to anything susceptible of carrying significance or meaning: images, sounds, and so on. In this way an absolute Knowledge (*Savoir*) can construct a mental space for itself, the connections between signs, words, things and concepts not differing from each other in any fundamental manner. Linguistics will thus have established a realm of certainty which can gradually extend its sovereignty to a good many other areas. (133)

Here, ultimately, everything, even space, is language.

Lefebvre contrasts this "optimistic" view with another, pessimistic one:

> An examination of signs reveals a terrible reality. Whether letters, words, images or sounds, signs are rigid, glacial, and abstract in a peculiarly menacing way.... Written, they serve authority. What are they? They are the doubles of things. When they assume the properties of things, when they pass for things, they have the power to move us emotionally, to cause frustrations, to engender neuroses. As replicas capable of disassembling the "beings" they replicate, they make possible the breaking and destruction of those beings, and hence also their reconstruction in different forms. (134)

And here, with some hyperbole, Lefebvre notes a very different view of language, in which one can do things with words, things that one could not do without them.

Now, according to Lefebvre, it is not common for these two notions of language to be presented as "pure types"; rather, most views of language tend to combine them together. Unfortunately, he continues, whatever the merits of this eclecticism, the result is typically a view that fails to deal adequately with space, this despite the fact that

> signifying processes (a signifying practice) occur in a space which cannot be reduced either to everyday discourse or to a language of texts.... [Indeed] the deployment of the energy of living bodies in space is forever going beyond the life and death instincts and harmonizing them. (136–37)

The solution to the problem of language, then, is to see the process of signifying as occurring *in* space: "Once brought back into conjunction with a (spatial and signifying) social practice, the concept of space can take on its full meaning. Space thus rejoins material production" (133). In fact, when we begin to see language as a set of social practices, we can begin to see, after Nietzsche, that metaphor and metonymy are not figures of speech but acts. As Lefebvre puts it, quoting Nietzsche, language is "a mobile army of metaphors, metonyms, and anthropomorphisms—in short, a sum of human relations, which have been enhanced, transposed, and embellished poetically and rhetorically, and which after long use seem firm, canonical, and obligatory to people."[8] Metaphor and metonymy, therefore, "decode, bringing forth from the depths not what is there but what is sayable, what is susceptible of figuration—in short, language" (139). And they do this in the form of actions that are creative.

At first glance it may appear that Lefebvre has offered a view of language that gets beyond the traditional view of language as a system of signs mappable onto the world. Indeed, that modernist view seems in important respects to be like the first view Lefebvre described, the view that he associated with linguistics. And, too, he appears to have moved beyond the second pessimistic view, doing so through his focus on the situatedness of language within social space.

Yet, it seems to me that he has left out some important questions here, and that these questions, in fact, concern the implicit spatiality of his conception of language, and its relationship with ideas, texts, and the world. Lefebvre's work brings up the issue in the English-speaking world simply by existing in English, for it was, of course, written in French. The fact that it has been translated is hardly kept a secret; indeed, the translator's name is predominantly featured on the cover. And the publisher claims that the work is characterized by "a deftness of style which Donald Nicholson-Smith's sensitive translation precisely captures." Now if there is agreement that this is a good translation, there is no real indication in the work (which was not the first of Lefebvre's to be translated) about what, exactly, it means for a work to be a good translation, or even what it means to be "translated" at all.

In one sense this may seem like a quibble. After all, translation is an accepted practice. And yet, the possibility of translation points inevitably to underlying assumptions about the nature of texts, about what their constituents are, and about what it means for two people to "say the same thing," and under what conditions that can be said to occur. Nicholson-Smith in fact alludes to the problem in a footnote. Speaking of the two French terms for the English "knowledge," he says,

> The *savoir/connaissance* distinction cannot be conveniently expressed in English. Its significance should be clear from the discussion here.... Wherever the needs of clarity seemed to call for it, I have indicated in parentheses whether "knowledge" renders *savoir* or *connaissance.* (11 n. 16)

The important point here is that the translator is *not* claiming that the distinction between two types of knowledge is one confined to the French, and that English speakers will need to grasp something quite new. On the contrary, he is claiming that there exist wholly outside of language two types of knowledge, and that the French have two words for them, whereas the more impoverished English-speaking world has tried to get by (to its detriment, it turns out) with one. But both the French and the English are using language for something that is nonlinguistic. Here language is viewed as a set of representations of nonlinguistic entities, including concepts and theoretical constructs.

This is a common enough idea and is difficult to get around. Even if we grant with the most radical of linguistic relativists that there are some things that can be said in one language and not in another, few among us would be likely to claim that it makes any difference at all to a mathematician whether one uses the word "two" or the word "deux"; both seem to refer to the same concept. But my point here is rather different. It is that if we consider the various ways in which the notions of "translatability" and "sameness of meaning" are typically conceptualized, those ways inevitably appeal to notions of space. For example, one traditional way of thinking about the meaning of words is in terms of reference; here, two words can be said to mean the same thing if they have the same extension, that is, if they refer to the same set of objects. But here the theory of meaning relies on the notion of a "set of objects." And this, in turn, suggests a range of images, all spatial, of the objects together in a room, or surrounded by some sort of barrier, or, alternatively, of the word pointing to the objects, picking them out.

By contrast, a second way of thinking about sameness of meaning sees meaning as a matter of criteria; here, a term is defined in terms of some set of criteria, characteristics, or essences. For two words to have the same meaning is for them to be defined in terms of the same criteria. This, of course, is a traditional way of thinking about meaning within the sciences, where a human, gene, or process is defined in terms of a set of attributes. Here too, however, meaning appeals to implicitly spatial notions, in this case to a notion of conceptual space; that is, there is a tendency to operate in terms of images like Venn diagrams, seeing meeting or not meeting criteria as a matter of being inside or outside of a particular conceptual space, and seeing complex definitions as matters of intersections within conceptual space.

Both of these views of meaning, the extensional and the intensional, are quite modern in character; indeed, they are the very ones that Rorty and others have roundly criticized. They replaced earlier notions of language, where words were taken to be natural signs, sharing something essential with that to which they referred. The appeal to natural signs is surely a way to develop a nonspatial way of thinking about language, but it is one on which few are willing to take a chance.

But more recently, spatial theories of meaning have been replaced, or at least augmented, by an alternative way of thinking about language and about sameness of meaning, which in a sense skirts this appeal to spatial imagery. This alternative, which Lefebvre takes, moves the concern from representation to practice, and he adopts this approach, as I noted earlier, repeatedly, and particularly in his discussions of the nature and development of metaphor. In the place of the concern with meaning, it puts a concern with the effects of language, with pragmatics. Here language is seen as a set of practices that do things. One can then take expressions of spatial assumptions less as expressions of beliefs in some spatial structure than merely as individual actions. When looked at from that perspective, a set of statements by an individual need not refer back to some underlying set of beliefs, they need not be derivable syllogistically from those beliefs or assumptions but may instead be seen simply as elements of a linguistic repertory.

There are, though, problems with this approach to language, and not the least of them is that it presents difficulties when one attempts to distinguish between language and noise. If we think of language simply as one of many human practices, then on what grounds can we make a distinction between language proper and other practices? Lefebvre, it turns out, does recognize this problem, as where he claims that "signifying processes (a signifying practice) [are] . . . forever going beyond the life and death instincts and harmonizing them" (136–37). And he has something of an answer to this question, as where he claims that metaphor and metonymy "decode, bringing forth from the depths not what is there but what is sayable, what is susceptible of figuration—in short, language" (139).

Yet this response, this characterization of language as an emergent phenomenon, itself relies on another spatial metaphor, on a distinction between that which is deep, hidden and that which is revealed. And this spatial metaphor is absolutely central to Lefebvre's work.

In fact, it arises in the very formulation of the work. He claims,

> The project I am outlining, however, does not aim to produce a (or *the*) discourse on space, but rather to expose the actual production of space by bringing the various kinds of space and the actual modalities of their genesis together within a single theory. (16)

Here the distinction between the surficial and the real, and the view that one process hides beneath another both suggest that the function of the social sciences is to move below the surface, to look underneath in order to reveal what is "really happening." In a sense this view is the complement of the normal view of theory in science; if we think of theories as conceptual structures that may be laid like a net over reality, here reality itself can be seen as containing those theories, which are hidden beneath the surface, there to be revealed through the use of the right techniques.

It would, actually, be possible to develop an account of the production of space that did not rest on these notions of theory and fact, depth and surface. I have in mind Lefebvre's discussion of language more generally, and especially of metaphor. But where Lefebvre tries to do so, he relies once again on a spatial view. This is because his appeal to the place of metaphor in creative practice implicitly assumes that there is something that is *non* metaphoric, that is, some meaning that is stable and literal. And with that assumption Lefebvre is right where he started, with notions of denotation and connotation, extension and intension.

This leads to a final point. I have noted that in several ways Lefebvre appeals implicitly to spatial images and metaphors in his conceptualization of the relationships among language, ideas and meanings, and the world. This appeal happens in one way when he allows translation and proclaims metaphor and suggests an intermediary world of stable ideas. It happens in another way as he appeals to a distinction between the hidden and the surficial, where order is at once above and below reality. But both of these views come together in his view of the ways in which the ideas expressed in different works or by different people can themselves "come together."

We are all familiar with, to the extent that it seems an unquestioned commonplace, the fact that we live in a world full of books. Equally commonplace is the fact that those books are not unrelated, that it is possible to see connections among them. These are not physical connections, and as the beginning student of intellectual history or cultural anthropology or physics learns, they are not simply matters of the use of identical words. Rather, they are usually taken to be matters of an author "saying the same thing" in different texts, or multiple authors "talking about the same thing" in their own texts.

Now, in the sciences there has long been a view, one that I have associated with Descartes, that individual scientific works may come together into a kind of supratext, which integrates the ideas of individual texts into one large virtual text, which is the expression of the nature of reality. The image of such a text is perhaps most often appealed to both as ideal and as real in physics, where the twentieth century has seen a concerted effort to unite the understanding of gravity, electricity, and magnetism under a single conceptual scheme, and where the failure to do so has not dimmed the view that the actual practice of physics involves at a smaller scale the unification of ideas.

As I have noted, this view of science has for the last thirty years been subjected to a stream of criticism, as by Thomas Kuhn, who has argued that different periods or subdisciplines in science operate in terms of conceptions of what is normal, expressed in sets of practices acquired in training and education. On this view, it may very well be that the concepts used by different groups, while appearing the same, are not only different but incommensurable.[9] Similarly, in the same period Arthur Danto offered a critique of this view of history, claiming that its assumption of the possibility of a kind of suprahistorical document failed to see the extent to which the development of narratives situated an author in a particular time and place.[10] More recently, of course, the criticisms of social constructivist students of science and in history of people like Hayden White have developed related critiques.[11]

Yet, although these works have all called into question the theoretical underpinnings of the idea of the virtual supratext, and even called into question the possibility of theoretical underpinnings, they have failed to make any obvious changes in the ways in which science is actually practiced. The reasons for this failure are complex, and they extend beyond issues of representation into the issues of social, economic, and legal structures and practices that I shall discuss in the final two sections of this chapter. But here I would point once again to a simple but powerful feature of scientific texts—the fact that by and large widely different individuals and groups believe themselves able to understand them and to grasp "what the author meant," even when the texts have been rendered into another language. By and large, people see themselves as arguing about ideas and not about words on a page or about individuals.

And it is these facts that offer compelling everyday support to the notion that ideas exist apart from language, and that it is the ideas that are the real content of a scientific text. Because these ideas appear to have qualities—stability, permanence, clarity—that are greatly valued and that set them apart from the flux of everyday life, they are viewed as separate, as existing on a higher plane. It is the pervasiveness of this imagery, and the extent to which it is multiply supported, that renders ineffectual Lefebvre's attempts to evade its effects.

The Work in the World

PATTERNS OF AUTHORITY

If we turn away from issues associated exclusively with knowledge, language, and ideas, and away from issues concerned with authorship, reading, and representation, there remain other ways in which Lefebvre has invoked unexamined notions of space in his text. One of those is in the matter of social organization and authority.

Explicitly, at least, he is attempting to establish an understanding of society, and here of space, that will in some way be liberating, that will help people to cast off the burdens of a repressive and unfair social and economic order. Indeed, he ends the volume with a hopeful chapter titled "Openings and Conclusions," which is unlike normal "conclusions" in the way it points toward the future. Yet despite these hopes, there is a fundamental sense in which his work fits nicely—notwithstanding his desires—into a set of social systems, and a set of social systems that are commonly conceived of in spatial terms.

Lefebvre's work, as I have suggested earlier, looks like other works in geography, the social sciences, and the sciences. It has chapters, sections, footnotes, and an index. Each of these is a means for making explicit the author's judgments about relevance and importance, but each, too, does more. Perhaps most important here are the notes and references, for they are the means by which the author does two things. He establishes himself within an intellectual space (which is a social space), and within that space he positions himself hierarchically.

The hierarchical organization of science is, of course, one of its fundamental features and is one on which I commented earlier in my discussion of the sociology of science of Robert K. Merton.[12] There I noted that the practice of citing both implicitly and explicitly establishes a social hierarchy. And as a work that both cites (and fails to cite) and is cited, Lefebvre's is therefore legitimately conceived of as operating within just such a hierarchy.

Lefebvre makes four kinds of citations. He cites great philosophical authorities, especially Hegel and Marx but also Descartes, Spinoza, Heidegger, and Nietzsche. He cites historical sources, like Viollet-le-Duc, Rabelais, and Vitruvius. He cites recent (recalling that the French version was published in 1974) authors who have written about space, among them E. T. Hall, Alexander Koyré, Erwin Panofsky, Christian Norberg-Schulz, and Christopher

Alexander. And he cites artists and social critics, like Picasso, Norman Mailer, and Octavio Paz.

That he cites these people is not at all unusual or unexpected, and it is that which may make it difficult to see what is happening here. If we look at the way in which he makes his citations, though, the social functions of those citations begin to emerge. The "great thinkers" whom he cites are all examples of what Foucault called "authors," those whom he believed to have created sets of ideas that have defined entire intellectual movements.[13] In fact, Foucault himself characterized these authors as people who defined an entire intellectual *space*, setting in place a way of viewing the world that competed with other, similar great sets of ideas. Their ideas come to constitute a kind of repertory of tools and questions. In citing Hegel, Marx, and Nietzsche positively, Descartes and Spinoza critically, Lefebvre situates himself in that intellectual space and at the same time makes it clear that what he is offering is an analysis of space that legitimates the subject through his association with men who are of lasting importance.[14] In legitimating the subject, he places himself in a space of authority.

A second set of references are to historical figures like Viollet-le-Duc and Vitruvius. Lefebvre's work, in fact, contains no original historical or social-scientific research; it relies, at best, on the original work of others, who are seen as evidence of the social thought of a time. Those authors were in their times writing controversial and theoretical works, but today their works are taken in a very different sense. And so, they are no longer subject to criticism but only to literary and historical analysis.

A third set of references are to theorists who at the time were actively pursuing the study of space. Here, in fact, virtually all of the references are negative in form. Lefebvre takes issue with Hall and Koyré and Panofsky and Maurice Merleau-Ponty, and on and on. The impression that we are given is that (in 1974) there has been no serious thought about space, that those who have thought about the matter have failed to grasp the issue—in part, of course, because they have not aligned themselves properly with the greats. Indeed, this view of current works is heightened by the cursory attention that he pays to these authors; rarely does one get the impression that he thinks it worthwhile to devote time to their analyses.

There is, actually, another side of the matter, and it is Lefebvre's silence about a substantial number of works available at that time that in various ways did indeed grapple with the issues that he was considering. He was conspicuously silent about James Blaut, William Bunge, Anne Buttimer, Walter Christaller, Richard Hartshorne, David Harvey, J. B. Jackson, Edward

Relph, Yi-Fu Tuan, and Edward Ullman, all of whom had written (sometimes substantial) widely available works. Both of these practices, his negative references and his silences, are additional means by which Lefebvre situates himself in a social space, in this case a social space where he is the only person doing serious work *on* space.

These three sets of practices together—citing greats, appealing to unquestionable historical sources, and dismissing the works of others on the subject—are traditional ways of establishing oneself within a subdiscipline and establishing oneself in a dominant position there and in reference to other subdisciplines. And these issues are almost inevitably—this is true in Lefebvre—conceptualized in spatial terms.

In the first instance, what has not been questioned, and what Lefebvre's organization of the text supports, is the notion that there *is* a hierarchy, which extends beyond the realm of "pure ideas" yet is implicitly expressed in the presentation and organization of those ideas. There is a second, and related, way in which the practice of citing has been associated with spatial conceptions, and that is in the matter of the characterization of the relationships among individuals within subdisciplines and of the relationships among subdisciplines. The development of subdisciplinary groups was characterized by Diana Crane as a matter of the creation of "invisible colleges."[15] This view, of a subdiscipline both as spatially dispersed and as very much like a traditional agglomerated institution, has through the use of citation analysis been formalized.[16] And a fundamental way of analyzing subdisciplines has relied on the use of cluster analysis, wherein the image of a subdiscipline is as a group of people who are "closer" to one another than they are to others, and a discipline, in turn, as a group of subdisciplines that are also closer to one another than to those outside. The notion of "closeness" here relies mathematically on the measurement of distances in a multidimensional space.[17]

Now, Lefebvre was not engaging in citation analysis or in the use of numerical taxonomy. But there are several important things to see here. First, the development of mathematical methods of citation analysis did *not* involve a reconceptualization of the nature of citations; rather, it simply translated traditional scholarly methods of textual analysis into quantitative terms, and in doing so made their presumptions all the more clear. And perhaps more important, the place of texts within the social system of science and of publishing makes these relationships all but inescapable. Although it is possible in some disciplines, like philosophy, to write without citing, it is all but impossible elsewhere; to attempt to do so is to risk being marginalized, to risk being placed, both literally and figuratively, at the margins of one's discipline.

Here it is important to note that we have been looking only at one side of the matter, at the references made by Lefebvre within his work. Now, it may be commonplace, but it is still incorrect, to separate Lefebvre's work from the references that have been made *to* it. One may wish to object that the references made to one's work are a matter utterly out of one's control, that even while an author is alive, he or she tosses a work into the world—and then it is on its own. But to adopt this view is to fail to see that any text is a piece of technology, which has a distinct set of purposes: to convince the reader of the correctness of a set of views, to convince the reader of the breadth or depth of understanding of the author, and so on.

Whereas at various times engaging in conversation or correspondence was the primary means for achieving these ends, in the twentieth century the mark of success has become the citation. And so, at the very least the possibility of one's being cited is built into the process of writing and publishing. But, in fact, by writing in particular ways one establishes the possibility of being cited in particular ways. In the case of Lefebvre, the ways in which he wrote in fact have engendered a particular pattern of citation, and it turns out to be one very much like the pattern that he used; that is, a pattern that at once establishes a subdisciplinary space and establishes a hierarchy within that space.

Between 1981 and 1992, the Social Science Citation Index reports 125 references to any of the works by Lefebvre. Of that total, 32 were to *The Production of Space,* and 44 were in journals of geography and planning. There are two striking features to these citations. First, they by and large appear within works of theory. In fact, of the 44 references to Lefebvre in geography and planning journals, only 7 have what might be termed substantial empirical content.[18] The overwhelming majority are to works of theory, to programmatic statements, and to reviews.

And second, the locations of citations are strongly skewed. References to Lefebvre's works appear in 76 journals, but over the twelve-year period, in only 20 journals more than once. Moreover, 34, or about 25 percent, of the citations appear in a total of only 5 journals. In fact, the journal in which the most references to Lefebvre occurs, *Environment and Planning D: Society and Space,* accounted for 18 citations, while the next highest journals (*International Social Science Journal, International Journal of Urban and Regional Research, Cahiers Internationaux de Sociologie,* and the *Annals of the Association of American Geographers*) had only 4 each.

In an important sense, then, *Society and Space* emerges as a place where a substantial debate about the nature of space is occurring. At the same time, there are certain unusual features about Lefebvre's treatment within the journal, including the publication of an interview and an obituary, which make him clearly someone worthy of special esteem.[19] Second, the quality of the work of those doing the citing, where he is taken as an intellectual authority by those whose theoretical orientations appear long since to have been established (as in the case of E. W. Soja and Allen Pred), indicates that he is someone of special persuasiveness.[20] And finally, his place in works widely seen as of importance in a range of debates, from flexible specialization to war to tourism, indicates that his own work needs to be seen as of unusual breadth and scope.[21] And so, we see in the pattern of references to Lefebvre the establishment of an ideational space, itself focused around a series of texts (in this case, a journal), and at the same time a hierarchical space, where the informing of current work by Lefebvre's work is a matter of deference.

ON THE WRITTEN WORK AS OBJECT

Even where there is agreement that a work like Lefebvre's ought to be seen as implicitly expressing spatially informed attitudes about both representation and social status, there is a tendency to fail to see that such a work also incorporates spatial notions to the extent that it is a book, and a book of a particular kind. As I have noted, books are tied to conceptions of space because they are elements in a series of systems, of classification and property, which are typically composed of elements that are characterized in spatial terms. To choose to communicate in the form of a book, rather than of some other medium, is already to implicate oneself in a set of systems and is by implication to legitimate those systems and their spatial implications.

As Lefebvre knows better than most, books are pieces of property, elements of a system of commodity circulation. International trade in books is, of course, regulated as a part of international trade more generally. But as I have suggested earlier, one thing that distinguishes both books and articles is that they are regulated as forms of intellectual property and as such fall within a highly developed system of regulations, the locus of which is the nation-state. Hence, just by engaging in the act of publishing a book, Lefebvre associates himself with a national-international spatial order.

Within this order his work is, from a legal standpoint, first and foremost a French work; that is, it gains its legal status in terms of French law. And from that initial point it becomes a part of the larger system. In fact, though, that it was initially a part of the *French* system of intellectual property rights, that it was written and published in France, gives it a special place within that system, one that sees individual texts as visible extensions of the author's personality.[22]

That they are pieces of intellectual property makes books special in several ways. One of those is that the author and publisher retain certain rights after the book has been sold. Under the French system of moral right, the author's right to control of the text does not expire and cannot be alienated. But even under that most restrictive regime, the publisher retains certain sorts of rights. There is, of course, the right to limit and require payment for quotation and reproduction. But there is also a right to limit the ways and forms in which the work may be circulated. In the case of Lefebvre, it is stipulated that

> except in the United States of America, this book is sold subject to the condition that it shall not, by way of trade or otherwise, be lent, re-sold, hired out, or otherwise circulated without the publisher's prior consent in any form of binding or cover other than that in which it is published and without a similar condition being imposed on the subsequent purchaser. (iv)

Thus the form in which the volume may be circulated, even circulated without payment, is regulated, and this circulation occurs within a highly organized system. In one sense, this is a physical system; books are circulated within libraries and among libraries. There are local, regional, and national libraries, just as there are ones associated with government, universities, and businesses. The purchasing and lending practices of these libraries are highly differentiated, and to choose to write a certain sort of book, to write for a certain audience, to write in a certain language, to choose a particular publisher all act to place one in a particular way within these systems.

But it is also a typological system; it requires that holdings be cataloged. Whether Lefebvre liked it or not, his work was, simply because of the way that he wrote and published it, made a part of several cataloging systems.[23] These ordain that in every library that uses a given system, a work be placed in a certain place. Under the Library of Congress system, for example, *The Production of Space* is located not with Yi-Fu Tuan's *Space and Place* and Robert Sack's *Conceptions of Space in Social Thought* (both G 71.5), or Richard Hartshorne's *Perspective on the Nature of Geography* (G 70), or even Gaston Bachelard's *Poetics of Space* (B 2430 B253P63), but rather with Mark

Heller's *The Ontology of Physical Objects: Four-Dimensional Hunks of Matter* and I. Rice Periera's *The Nature of Space,* an art volume published by the Corcoran Gallery (BD 621).

The Space in *The Production of Space*

By now it may seem as though we have a bewildering array of spatial conceptions, all at work in *The Production of Space.* We have seen the neutral author pitted against the author as adversary, and both against the author who is simply absent. We have seen the text as a piece in a hierarchy of references, the text as a stock of ideas, and the ideas as elements in a neutral conceptual space. We have seen the universal reader, outside of time and space, and the very particular reader, the object of enough training to be able to make sense of the volume. We have seen the relationship between the language of the work and the ideas it expresses laid out in different ways; language here maps onto ideas in a neutral, Platonic universe, while it there emerges from the world. We have seen the work as an element in a social hierarchy of science but also as part of a network of the citers and the cited. We have seen it as the work of an author both a Lockean individual and a Hegelian persona-in-the-world. And finally, we have seen the work as an element in a larger classificatory scheme, one arranged in a way that is decidedly hierarchical, that puts the work "in its place."

This is a long list, but it nonetheless remains possible to pare it down—although in the space that I have I shall be able to do so only in a summary way. When we do, we find in several places the consistent appeal to conceptions of hierarchy; we find them in the ways in which the work is socially situated and in the ways in which knowledge is conceptualized, as based on foundations represented by references.

If this much is unambiguously stated, much else is not. Whereas the process of citation appears to support a hierarchical view of space, the characterization of those who cite is more in terms of people who are close and far from one's position, in a way that suggests a kind of Leibnizian, relational space, and this characterization is amplified to the extent that Lefebvre has implicated himself in systems of citation analysis.

Similarly, both the author and the individual physical work are represented as individuals within what we might term a Newtonian space. Both, that is, are represented as able to maneuver with ease in that space; indeed, nowhere is this more clear than in the way in which the author presents himself as having been able to say what he meant, to take pen to paper in a way that is unconstrained. This sense of absolute space is represented in another way, where the author represents himself as standing outside and looking in, where being outside is a matter of having gained some absolute toehold.

Yet the author also takes a position, and to the extent that others are represented as holding views that are to be expected on the basis of who or where they are, the author moves variously into Aristotelian and neo-Kantian imagery. And the reader, too, occupies various forms of space; the reader needs at times to be able to read from a particular position, the position of the reader, but at other times takes on a universal cast, as if adrift in a neutral nonspace.

Finally, when we turn to the ways in which Lefebvre uses language and ideas, we find a range of views. Here language maps onto ideas, here language emerges from below; in the first case we seem to have a neutral space, but in the second, space seems well ordered and directional.

Why is this complexity of spatial conceptions important? Why does it matter? It matters for the following reason: Lefebvre set out to develop a critical understanding of space. Although it may be tempting to imagine that he individually and independently developed a theory, and only later, only as a matter of happenstance "wrote it up" and published it, this is plainly not what happened. Rather, the "coming up with the theory" is only a part of the project with which we have been presented; the ideas arrive in a book, with a publisher, an audience, citations, and on and on. If it is tempting to imagine that these last are only contingent features of the project, a moment's reflection will show that far from that they are essential and intrinsic because his intent all along was to communicate his views to an audience, to make them public, to gain assent, to change society. Yet in taking that essential next step he has absorbed into his project a whole set of ways of thinking about space, which are there for each reader to see, and because Lefebvre has appealed to them, for every reader to believe. He may have wanted his readers to come away with a particular critical image of space, but because his book is a work in the world, it leads the reader, implicitly or explicitly, to accept certain other images, themselves far from critical.

CONCLUSION

Learning from the Place of the Work in the World

The conclusion drawn from an analysis of Lefebvre's *The Production of Space* needs now to be expanded. We saw in this analysis that the commonsense view that he, or anyone else, ought to be able just to offer up a new theory of space, one compelling in its clarity and power, and have it accepted and applied is nothing more than a comforting image. In practice, the attempt to develop a comprehensive theory of space fails not because of the inability of the author to do it right but for the more fundamental reason that like all authors, Lefebvre is emplaced. In the ways that he writes and in the very ways that he organizes his work, he is defining it as an object suitable to be read, discussed, even dismissed in very particular ways.

This is also true in the ways that he cites; he defines a group who can variously be seen as supporters and opponents, conventionally thought of as themselves laid out in a conceptual space but perhaps better seen as occupying, or excluded from, potential places of dialogue. Moreover, and certainly he would not have known this when he was writing, his own citations are entangled in a larger network of citations, compiled and disseminated by computers, and now over computer networks.

Further, because he has chosen to make his ideas public by publishing them in the form of a book, he has tied himself to another set of places, to the bookseller and the publishing house. His work is a piece of property, one that can be quoted in certain ways by certain people, and not by others.

As an author in a system of intellectual property, he comes to be a part of a commodity system and a legal system, each of which defines him as an individual whose actions are limited in particular ways, by where he lives and publishes, by the language that he speaks. His work can be bought and sold under certain conditions and not under others; it can be taken into some countries and not into others. As an author who has said certain things, he is welcome in certain places and not in others.

He is tied into a system of academic status, where to be associated with him or his work places others, too, within that system, and where the system is widely *seen* as a system, as a place organized hierarchically. And as a book it is enmeshed in various systems of classification. Through them it is defined, and for all time, as like these works and not like those, again in a system that is fundamentally hierarchical.

I have argued that in all of these ways the individual choices that Lefebvre and others make as writers, academics, and authors involve them in a wide range of practices that make places of certain sorts and not of others, that enable them to do certain things and not others, and that exclude others from doing some things while otherwise enabling them. It is the fact that everyone who is involved in such work is engaged in this practice of place making that renders so problematic the attempt to get outside the system, to offer up a real alternative. For at every turn the theory is confronted by the practice, and at every turn the practice undercuts the theory.

I suggested at the outset that a consideration of the written work can offer some help in dealing with what often seem difficult and even intractable problems. Might I, as certain postmodernists have suggested, adopt a position of radical relativism, where each person can truly know only what he or she knows? If I decide that what I have been saying is wrong, can I not just change my mind? And, why is the written work so easily misunderstood?

Given that these are central questions, the reader who has gotten this far may be wondering what the answers are, and why they have been so long in coming. And in fact I have not spoken explicitly about any of the questions—although I certainly have laid out elements of an answer, and what I have just said about Lefebvre ought to be more than suggestive. And there is a reason for that, which ought also to have by now become apparent.

On the Possibility of Radical Relativism

The radical relativist claims that some features of a person's relationship with the world, a person's language or knowledge or even objects, are relative to that own person, perhaps to that person's culture or economic position or gender or race. And as a result, it is claimed, it is as though we live, each individual or group, cut off one from the other in a necessary and irremediable way. This is a view both disturbing and compelling. It is disturbing because it seems bleak, hopeless. And it is compelling because it seems so inexorably to follow from premises, even from everyday experiences, that seem so patently right.

Conceptually, at least, this view derives from two sources that have also figured in discussions about the nature of space. First, there is Kant, and his enunciation in the *Critique of Pure Reason* of a view in which we confront a world that we have already in some way constructed. Although for Kant this created no problems, because we all construct our worlds in the same way, for people in the twentieth century, confronted with a vaster and more diverse world, it has seemed to tell a different lesson. It seems a simple matter to draw from this view a form of radical relativism. Indeed, we are all, at least now and again, Kantians in this way, when we imagine that our enemies or neighbors, children or parents cannot understand us because they "live in different worlds."

In another sense this view derives, perhaps more unfairly, from Descartes and from modernism more generally. And here I mean three things. It derives from the view that it is possible to see the world as a system; it derives from the view that one can manipulate ideas in a sort of logical space, moving from premise inexorably to conclusion; and it derives from the view that we have minds that are invisible containers for our ideas. Whereas the Kantian element of human experience tells us that we *do* live in different worlds, the Cartesian tells us that from that fact we can draw a series of conclusions. We must conclude that because our ideas are hidden, they cannot be known. We must conclude that because we can imagine the world as a system, it is possible to draw very general conclusions about that system. And we can reach these conclusions through logic, through a process of reasoning from premise to conclusion.

The real problem is that this way of countering seems to leave intact the very ways of thinking about the world that seem to have led to the conclusions. These conclusions, after all, are not going to be undercut by logical arguments, because those arguments seem to lack the force to topple the images that underpin them. The notion that "I see the world in my own way" is, after all, in our society supported every day in a great many ways; we are bombarded with arguments and claims that seem comprehensible *only* if we believe that those making them see the world differently. And we are deluged with images, again every day, that tell us that we *can* look at the entire world and develop a comprehensive understanding of it.

What does this have to do with the written work? Simply that the written work is an element that both derives from and supports the commonsense experiences to which the Cartesian and Kantian arguments defer. The work provides a powerful image of a repository of ideas not unlike the mind. The written work appears to be an expression of a point of view; it is said to contain "what its author believed," and almost every book or article contains something odd, different, even inexplicable. At the same time, in the way that books circulate, through systems of booksellers but also through libraries, they provide an image of the book and its contents as something mobile, something that could be anywhere.

It is a commonplace to reject the conclusions of radical relativists, of postmodernists, and of those who argue for the unknowability of the written work because they appear self-defeating. After all, it is argued, if there is no truth, how can you claim that it is true that there is no truth? I think it fair to say, though, that this argument tends to convince only those who are ready to be convinced and leaves the rest ready to regroup and have at it again.

The real difficulty with radical relativism is the very difficulty that Sandra Harding pointed to in the matter of essentialism. The problem is that it only makes sense to make radically relativist claims in very specific situations, in certain places. Although the *evidence* for radical relativism appears omnipresent, only certain sorts of people actually embrace it, and they do so only in certain places. We have all had the experience of arguing with someone, perhaps an intractable undergraduate, who claims that "What you say is just your point of view. It's all relative." Leaving aside, again, the self-defeating character of these claims, what is more important is that the student only makes those claims in certain places. They are made in a classroom, in a university, in an argument, in an academic paper. They are not made in the toolroom, in the hospital, in battle. They are not made when setting up a die grinder or a milling machine or an automobile emissions tester. And the fact that they are not made there is not a matter of the people involved being too busy or too distracted. The fact is that in those situations claims like "That's just your opinion" make no sense.

That the situatedness of such statements is not seen is ironic, because it is a commonplace that one can read in certain places and not others. Books, more than many objects, simply belong in certain places. Although we are used to seeing Bibles flaunted in churches and once were used to seeing Mao's little red book waved at demonstrations, books in our society are far more usually tied to times and places of contemplation. Throughout the Middle Ages reading was done aloud; today the person who reads aloud in a public space is looked upon with scorn and suspicion.

And so, although we have very clear notions of where the written work belongs, that it has its place, the written work itself has been in some measure responsible for the belief that places do not matter. And in that way it has been responsible for the view that knowledge can somehow be separated from everyday actions and places and viewed as a disembodied set of ideas. In this way it has been responsible for the emergence of the claim that ideas can all be relative. Ironically, the place-bound practices of reading and writing, citing and editing, publishing and circulating, all of which are constricted in multiple ways that cannot simply be wished away, tell us that wishing is all we need.

Why Am I Misread?

An anonymous reader of an earlier version of the previous chapter commented as follows:

> This is an obtuse, narrow-minded, and mean-spirited piece of work that, if published, will do great harm to the intellectual status of geography as a discipline, especially if read by others working in related fields outside geography....

I am unable here to provide for the inquisitive reader further details of the nature of this harm, because that reader continued by saying, "I regret that I do not have the time or energy to substantiate this evaluation...." But no matter, because the precise nature of those details is less important than the fact of the disagreement. Gratified that someone believed that I had this much power, I nonetheless felt that at least in some small way this reader had misunderstood my intent. (Actually, and to be more honest, it seemed to me that the reader had misunderstood every word that I had said.)

How can this happen? In a sense, this is the question of relativism again. But it points, I think, to rather a different aspect of this same question. For whereas the question of relativism seems to begin from an assertion of fundamental difference, my question when faced with this evaluation of my work was quite the opposite: How can anyone who approaches my work with an open mind and a scholarly approach possibly reach such a conclusion? And in fact, it seems clear from a reading of the evaluation that the person criticizing my work believed precisely the same thing. There was no breast beating, no "This is just my opinion."

And so the question here, more precisely, is, How can people who explicitly espouse the same values reach such different conclusions about written works? When put in this way, the question begins to appear a bit different from the question about relativism, and for the following reason: it begins to point to the extent to which the nonrepresentational function of the written work in geography or in science more generally outweighs the representational.

Here I am pointing back to the issues that I raised in chapter 7, and especially to the ways in which the work is used as a means of establishing community and of constructing and maintaining places. This happens in a variety of ways. In the matter of Lefebvre, I pointed out that his work is widely cited in one journal and rarely anywhere else; in an important sense that journal is a place in which Lefebvre and his acolytes come together. But the written work is in other ways an armature around which places are constructed. Works are the topics of seminars, conferences, and reading groups. They are collected in a place in bookstores—the Marx section—and in the library. More recently, one finds people interested in particular topics congregating in places in cyberspace.

In each of these places one learns that certain ways of talking are acceptable and that others are not, that certain responses are acceptable and others are not. In the places in which Lefebvre's followers congregate, for example, to mention that he cites only one woman in a four-hundred-page volume, or that he fails to cite the dozens of people who have written important works about space, or that his is not the sort of work that one carries to the barricades is very much like spitting in the soup. This is not to say that the places that I frequent are somehow more pure, less subject to the whims and prejudices and ambitions of their inhabitants. But I shall return to that matter.

Why Is It So Hard to Change One's Mind?

If one thing is clear about the state of the academy in recent years, it is that it is changing. In the 1950s few in geography could have predicted the rapidity of the quantitative revolution; in the 1960s few could have predicted the rapidity of its fall from grace. It seems fair to say that few would have predicted that in the 1990s the once moribund field of industrial location would become a hot area, and that fewer would have predicted the entry of queer studies into the discipline. It seems equally true that to many outsiders the academy is, increasingly, a place in which the irrelevant is studied in offensive ways and represented in prose that is utterly inscrutable.

All of that being the case, it may seem odd that I raise the question, Why is it so hard to change one's mind? And yet it strikes me that a broad view of the field of geography, and of other fields as well, leads one to the conclusion that in important ways academics very seldom change their minds.

In one sense, of course, the suggestion that we do not change our minds is quite false. Certainly in the course of writing this book I have changed my mind about a great many issues. And trivially, the geographer who today says, "The population of Los Angeles is increasingly poor," and tomorrow says, "The population of Los Angeles is increasingly wealthy" has had a change of mind. But in making this claim I really mean two things, one perhaps more commonplace than the other.

The more commonplace claim is this: what I believe is available only to the extent that I make it public. And to the extent that as an academic I am forced to operate within a limited set of public places, I am equally constrained in the set of beliefs that I can represent. This, of course, was a central point in the previous two chapters and was equally central in my discussion of relativism. In a classroom or discussion group or conference or article or book, I can only say certain sorts of things. My ability to say new or unusual things is limited in a number of ways. There are laws and institutional rules that proscribe my making certain sorts of public comments, about race or sex or drugs or politics, for example. There are sets of mechanisms that more strongly restrict my making similar claims in the classroom; among them are the evaluations that students make of my teaching.

When I turn my lecture notes into books and articles, I am faced with a different set of restrictions. Editors and reviewers, especially anonymous reviewers like the weary, busy one I quoted earlier, restrict both what I can say and how I can say it. In some cases there are further restrictions, where my work might be libelous or might constitute an infringement of a person's privacy.

And of course there are broader constrictions. Researchers whose work is considered too obscure or too banal are simply denied employment and grants; they are excised from the entire system of production of written works. To go too far is to risk being marginalized, to risk being seen as beyond the pale.

Given all of these factors that conspire to keep one from changing one's mind, one may wish here to make a Cartesian counter—that the public expression of my beliefs is not the "real me," but rather a persona, a kind of alter-ego. "I've changed my mind, but I don't let on." It seems to me that this sort of claim raises a deeper sort of question about which it is difficult to make definitive statements. Here I would return to my earlier discussion of Harvey's work and point to two ways in which it seems to me to show substantial continuity, even beyond the obvious changes. First, and he makes this explicit, he represents his career as a series of quests, of intellectual ventures. Whether in the positivist *Explanation in Geography* or in the Marxist/cultural *The Condition of Postmodernity,* this sense of an ongoing project remains. And second, there remains through his work what I would term, perhaps somewhat hazily, an "intellectual style." Part of this is reflected in this sense of one's work as a series of intellectual quests, and there are a number of other ways in which one might imagine that work. But another, important part of it lies in one's way of approaching problems, in one's tolerance for ambiguity and vagueness, in the ways that one has of shaping a set of issues into a finished product.

When I say that I am hazy in my use of the term "intellectual style," I mean in part that such styles are themselves hazy, to the extent that they are defined by the sets of everyday practices in which geographers and scientists and scholars engage.[1] One might want to see such a style as reducible in some way to a set of psychological traits, but it seems to me that one need not be reductive in this way. To have an intellectual style is to research and read and write and cite and argue in particular ways and in particular places.

On Reflexivity

Each of the questions that I have raised here has been related to the more specific issue of this book, the nature of the written work in geography, just as the book has been meant as a means to the elucidation of those issues. What I have said about each here has been meant to be merely suggestive; my intent in these last few pages has not been to provide definitive answers to these questions but rather to point to the ways in which one might sort through what I have said in an attempt to begin the construction of more explicit answers.

Some may find this approach maddening, especially in light of the following question: Does what you have said here apply to *this* work? This seems to me to be an important question, and putting matters most simply, my answer is this: Of course I take what I have said to apply to this work. But what I have said suggests strongly—indeed, I take the evidence to be overwhelming—that it is a mistake to imagine that one can construct a theoretical edifice that can somehow encompass and make sense of the world. Rather, *any* theoretical construction, to the extent that it involves doing research, writing, reading, citing, discussing, publishing, and so on, in a fundamental sense is a matter of the carrying on of sets of practices in places. The use of theory is itself a matter of practice, and as I have suggested, theories are most often used not as machines to generate answers but rather as images to bring some people into a fold and keep some out.

Some might suggest another approach, that I ought to be more explicit about my own interests, prejudices, desires. Yet to do so would itself be to invoke a theory, a theory of the role of interests, prejudices, and desires.

So with respect to the question of how what I have said can be said to apply to this and my other work, the answer is not that I believe myself somehow to have constructed a theory that encompasses and makes sense of all other works, rendering them objects in my grander scheme. Quite to the contrary, I take this work to have shown the impossibility of such a project and to have shown the need, instead, to put the geographical work in its place, to see the work in the world.

Notes

Introduction

1. Clifford Geertz, "Common Sense as a Cultural System," *Antioch Review* 33 (1975): 5–26.

2. Joseph Gusfield, "The Literary Rhetoric of Science: Comedy and Pathos in Drinking Driver Research," *American Sociological Review* 41 (1976): 16–34; Donald McCloskey, "The Rhetoric of Economics," *Journal of Economic Literature* 21 (1983): 481–517; Donald McCloskey, *The Rhetoric of Economics* (Madison: University of Wisconsin Press, 1985); John S. Nelson, Allan Megill, and Donald McCloskey, eds., *The Rhetoric of the Human Sciences: Language and Argument in Scholarship and Public Affairs* (Madison: University of Wisconsin Press, 1987).

3. See, of many, Charles Bazerman, *Shaping Written Knowledge: The Genre and Activity of the Experimental Article in Science* (Madison: University of Wisconsin Press, 1988); J. A. Campbell, "Scientific Revolution and the Grammar of Culture: The Case of Darwin's Origin," *Quarterly Journal of Speech* 72 (1986): 351–76; Alan G. Gross, *The Rhetoric of Science* (Cambridge: Harvard University Press, 1990); Greg Myers, *Writing Biology: Texts in the Social Construction of Scientific Knowledge* (Madison: University of Wisconsin Press, 1990); and Lawrence J. Prelli, *A Rhetoric of Science: Inventing Scientific Discourse* (Columbia: University of South Carolina Press, 1989).

4. Clifford Geertz, *The Interpretation of Cultures* (New York: Basic Books, 1973); Clifford Geertz, "Thick Description: Toward an Interpretive Theory of Culture," in *The Interpretation of Cultures,* pp. 3–30; George Marcus, "Rhetoric and the Ethnographic Genre in Anthropological Research," *Current Anthropology* 21 (1980): 507–10; George Marcus and Dick Cushman, "Ethnographies as Texts," *Annual Review of Anthropology* 11 (1982): 25–69; Clifford Geertz, "'From the Native's Point of View': On the Nature of Anthropological Understanding," in *Local Knowledge: Further Essays in Interpretive Anthropology* (New York: Basic Books, 1983), pp. 55–72; James Clifford, "On Ethnographic Self-Fashioning: Conrad and Malinowski," in *The Predicament of Culture: Twentieth-Century Ethnography, Literature, and Art* (Cambridge: Harvard University Press, 1988), pp. 92–113; James Clifford, "On Ethnographic Allegory," in James Clifford and George Marcus, eds., *Writing Culture: The Poetics and Politics of Ethnography* (Berkeley: University of California Press, 1986), pp. 98–121.

5. Pierre Bourdieu, *The Logic of Practice,* trans. Richard Nice (Cambridge: Polity, 1990); Michel de Certeau, *The Practice of Everyday Life,* trans. Steven Rendell (Berkeley: University of California Press, 1984); Anthony Giddens, *Central Problems in Social Theory* (Berkeley: University of California Press, 1979); Peter Winch, *The Idea of a Social Science and Its Relation to Philosophy,* 2d ed. (London: Routledge, 1990).

6. Stephen A. Tyler, "Post-Modern Ethnography: From the Document of the Occult to the Occult Document," in Clifford and Marcus, eds., *Writing Culture,* pp. 122–40; Stephen A. Tyler, *The Unspeakable: Discourse, Dialogue, and Rhetoric in the Postmodern World* (Madison: University of Wisconsin Press, 1987); Vincent Crapanzano, "The Postmodern Crisis: Discourse, Parody, Memory," in George E. Marcus, ed., *Rereading Cultural Anthropology* (Durham, N.C.: Duke University Press, 1992), pp. 87–102; Vincent Crapanzano, *Tuhami: Portrait of a Moroccan* (Chicago: University of Chicago Press, 1980).

1. Formalizing Common Sense (I): The Text as Representation

1. Richard Rorty, *Philosophy and the Mirror of Nature* (Princeton: Princeton University Press, 1979).

2. Jack Goody, *The Domestication of the Savage Mind* (Cambridge: Cambridge University Press, 1977); Eric A. Havelock, *The Muse Learns to Write: Reflections on Orality and Literacy from Antiquity to the Present* (New Haven: Yale University Press, 1986); Eric A. Havelock, *Preface to Plato* (Cambridge: Harvard University Press, 1963); Walter J. Ong, *Orality and Literacy: The Technologizing of the Word* (London: Routledge, 1982); and Walter J. Ong, "System, Space, and Renaissance Symbolism," *Study on Renaissance Symbolism* 8 (1956): 222–39.

3. Ong, "System, Space, and Renaissance Symbolism," p. 226.

4. Milman Parry, *The Making of Homeric Verse: The Collected Papers of Milman Parry,* ed. Adam Parry (Oxford: Clarendon Press, 1971).

5. Plato, *Phaedrus,* trans. R. Hackforth (Cambridge: Cambridge University Press, 1972). In the case of the works of Plato, as of others, page references are to the standard editions. Subsequent page references will be given in the text in parentheses.

6. Brian Vickers, *In Defence of Rhetoric* (Oxford: Clarendon Press, 1988), p. 113.

7. Ibid., p. 111.

8. Plato, *Sophist,* 254a. References are from "The Sophist," in *Plato's Theory of Knowledge,* trans. F. M. Cornford (London: Routledge and Kegan Paul, 1935), pp. 165–332. Subsequent page references will be given in the text in parentheses.

9. Plato, *Gorgias,* 462c; in *The Dialogues of Plato,* ed. R. E. Allen (New Haven: Yale University Press, 1984), pp. 187–316. Subsequent references will be given in the text in parentheses.

10. Plato, "Menexenus," 235a; in Allen, ed., *The Dialogues of Plato,* pp. 317–43.

11. Plato, *The Republic,* trans. F. M. Cornford (New York: Oxford University Press, 1945).

12. Plato, *Meno,* in Allen, ed., *The Dialogues of Plato,* pp. 131–86.

13. Plato, "Cratylus," 388; in *The Dialogues of Plato,* trans. Benjamin Jowett (New York: Oxford University Press, 1892), pp. 1–107.

14. Plato, "Cratylus," 439b.

15. Gunnar Olsson, "The Eye and the Index Finger: Bodily Means to Cultural Meaning," in Reginald G. Golledge, Helen Couclelis, and Peter Gould, eds., *A Ground for Common Search* (Goleta, Calif.: Santa Barbara Geographical Press, 1988), pp. 126–37; Gunnar Olsson, "The Social Space of Silence," *Poetica et Analytica* 3 (1985): 6–31; Gunnar Olsson, "Toward a Sermon of Modernity," in Mark Billinge, Derek Gregory, and Ron Martin, eds., *Recollections of a Revolution: Geography as a Spatial Science* (New York: St. Martin's Press, 1983), pp. 73–85; Allan R. Pred, "Lost Words as Reflections of Lost Worlds," in Golledge, Couclelis, and Gould, eds., *A Ground for Common Search,* pp. 138–47; Dagmar Reichert, "Comedia Geographica: An Absurd One-Act Play," *Environment and Planning D: Society and Space* 3 (1987): 335–42; Dagmar Reichert, "Writing Around Circularity and Self-Reference," in Golledge, Couclelis, and Gould, eds., *A Ground for Common Search,* pp. 101–25.

16. Michel Foucault, *The Order of Things: An Archaeology of the Human Sciences* (New York: Pantheon Books, 1971; reprint, New York: Vintage Books, 1973).

17. Martin Heidegger, "The Age of the World Picture," in *The Question Concerning Technology and Other Essays,* trans. William Lovitt (New York: Garland, 1977), pp. 115–54.

18. Stephen Toulmin, *Cosmopolis: The Hidden Agenda of Modernity* (New York: Free Press, 1990).

19. Dalia Judovitz, "Representation and Its Limits in Descartes," in Hugh J. Silverman and Donn Welton, eds., *Postmodernism and Continental Philosophy* (Albany: State University of New York Press, 1988), pp. 68–84. See also, Dalia Judovitz, *Subjectivity and Representation in Descartes: The Origins of Modernity* (Cambridge: Cambridge University Press, 1988).

20. Hans Aarsleff, *From Locke to Saussure: Essays on the Study of Language and Intellectual History* (Minneapolis: University of Minnesota Press, 1982); James Knowlson, *Universal Language Schemes in England and France* (Toronto: University of Toronto Press, 1975); Mary Slaughter, *Universal Languages and Scientific Taxonomy in the Seventeenth Century* (Cambridge: Cambridge University Press, 1982).

21. Francis Bacon, "The New Organon," in *Works,* ed. James Spedding, Robert Leslie Ellis, and Douglas Denon Heath (Boston: Taggard and Thompson, 1863), pp. 57–350. References are to paragraph numbers. Subsequent citations will be given in the text in parentheses.

22. Note that in *The Emergence of Probability* (Cambridge: Cambridge University Press, 1975), Ian Hacking argues that what Bacon calls induction might more properly be seen as what Peirce called "abduction."

23. Francis Bacon, "The Masculine Birth of Time," in *The Philosophy of Francis Bacon: An Essay on Its Development from 1603 to 1609, with New Translations of Fundamental Texts,* ed. Benjamin Farrington (Liverpool: Liverpool University Press, 1970), pp. 62–63.

24. Bacon, "The Masculine Birth of Time," pp. 59–72.

25. Paolo Rossi, *Francis Bacon: From Magic to Science,* trans. Sacha Rabinovitch (London: Routledge & Kegan Paul, 1968), p. 16. Much of what I shall have to say about Bacon and memory draws upon Rossi's work.

26. Bacon, *Novum Organon,* II; quoted in Rossi, *Francis Bacon: From Magic to Science,* pp. 202–3.

27. Cicero, *De Oratore,* Book II, ch. LXXXVI; in *On Oratory and Orators,* trans. and ed. J. S. Watson (Carbondale, Ill.: Southern Illinois University Press, 1970), p. 186.

28. Frances Amelia Yates, *The Art of Memory* (Chicago: University of Chicago Press, 1966).

29. René Descartes, *Descartes: Philosophical Letters,* ed. Anthony Kenny (Oxford: Clarendon Press, 1970). In referring to Descartes's work, I have used the following editions: for *Discourse on the Method* (1637) and *Meditations on First Philosophy* (1641), I have used *Descartes's Philosophical Writings,* ed. G. E. M. Anscombe and Peter Geach (Indianapolis: Bobbs-Merrill, 1971). For *Rules for the Direction of the Mind* (1628) and the *Optics* (1637), I have used *Selected Philosophical Writings,* trans. John Cottingham, Robert Stoothoff, and Dugald Murdoch (Cambridge: Cambridge University Press, 1988), pp. 57–72. In each case I have cited page references from the standard edition of Adams and Tannery, and for the *Philosophical Letters,* I have used the version cited above. Subsequent citations will be given in the text in parentheses.

30. Descartes, Letter to Mersenne, 20 Nov. 1629.

31. Descartes, *Oeuvres,* X, 230; quoted in Yates, *The Art of Memory,* p. 373.

32. Judovitz, "Representation and Its Limits in Descartes," pp. 69–72.

33. Slaughter, *Universal Languages and Scientific Taxonomy in the Seventeenth Century,* p. 127.

34. David Harvey, *Explanation in Geography* (London: Edward Arnold, 1969).

35. See Peter Gould, "Reflective Distanciation through Metamethodological Perspective," *Environment and Planning B,* 10 (1983): 381–92; and for works by other geographers, see Helen Couclelis, "On Some Problems of Defining Sets for Q-analysis," *Environment and Planning B,* 10 (1983): 423–38; and J. Johnson, "Some Structures and Notation of Q-analysis," *Environment and Planning B,* 8 (1981): 73–86. For the background to this work, see R. H. Atkin, "An Approach to Structure in Architectural and Urban Design. 1. Introduction and Mathematical Theory," *Environment and Planning B,* 1 (1974): 51–67; R. H. Atkin, "An Approach to Structure in Architectural and Urban Design. 3. Illustrative Examples," *Environment and Planning B,* 2 (1975): 21–57; R. H. Atkin, *Mathematical Structure in Human Affairs* (New York: Crane, Rusak, and Co., 1974); and R. H. Atkin, *Multidimensional Man* (Harmondsworth: Penguin, 1981).

36. Michel de Montaigne, *Montaigne's Essays,* trans. J. M. Cohen (Harmondsworth: Penguin, 1958).

37. Judovitz, *Subjectivity and Representation in Descartes,* p. 13.

38. Jaakko Hintikka, "Cogito, Ergo Sum: Inference or Performance," *Philosophical Review* 71 (1962): 3–32; John Austin, *How to Do Things with Words,* 2d ed., ed. J. O. Urmson and Marina Sbisa (Cambridge: Harvard University Press, 1975).

39. Hacking, *The Emergence of Probability,* p. 31.

2. Formalizing Common Sense (II): The Work in the System

1. David A. Kronick, *A History of Scientific and Technical Periodicals: The Origins and Development of the Scientific and Technological Press, 1665–1790* (New York: Scarecrow Press, 1962), p. 51.

2. Douglas McKie, "The Scientific Periodical from 1665 to 1798," in A. J. Meadows, ed., *The Scientific Journal* (London: Aslib, 1979), p. 123; see also Martha Ornstein, *The Role of Scientific Societies in the Seventeenth Century* (Hamden: Archon Books, 1963).

3. Derek J. de Solla Price, *Little Science, Big Science* (New York: Columbia University Press, 1963), pp. 6–18.

4. Ludwik Fleck, *Genesis and Development of a Scientific Fact* (Chicago: University of Chicago Press, 1979).

5. Robert K. Merton, "The Normative Structure of Science," in *The Sociology of Science: Theoretical and Empirical Investigations,* ed. Norman W. Storer (Chicago: University of Chicago Press, 1973), pp. 267–78; Robert K. Merton, *Science, Technology, and Society in Seventeenth-Century England* (New York: Harper and Row, 1970).

6. Merton, "The Normative Structure of Science," p. 273.

7. Nico Stehr, "The Ethos of Science Revisited: Social and Cognitive Norms," *Sociological Inquiry* 48 (1978): 172–96; Richard Wunderlich, "The Scientific Ethos: A Clarification," *British Journal of Sociology* 25 (1974): 373–77.

8. Merton himself has made this argument, in Robert K. Merton, "The Ambivalence of Scientists," in Storer, *The Sociology of Science,* pp. 383–412; see also, for example, Ian I. Mitroff, *The Subjective Side of Science: A Philosophical Inquiry into the Psychology of the Apollo Moon Scientists* (Amsterdam: Elsevier, 1974).

9. Michael J. Mulkay, "Norms and Ideology in Science," *Information sur les Sciences Sociales* 15 (1976): 637–65.

10. Ernest Cushing Richardson, *Classification: Theoretical and Practical* (New York: H. W. Wilson, 1930), p. 52.

11. Richardson provides a 110-page listing of such systems. Briefer accounts can be found in Arthur Maltby, *Sayers' Manual of Classification for Librarians,* 5th ed. (London: Andre Deutsch, 1975); and W. C. Berwick Sayers, *An Introduction to Library Classification,* 6th ed. (London: Grafton, 1943).

12. Of course, there is the earlier "French System," or System of the Paris Booksellers, the basis of which was established in the seventeenth century, and which was codified in the early nineteenth by Jacques Charles Brunet. See Maltby, *Sayer's Manual,* p. 111.

13. Christopher Merrett, *Map Cataloguing and Classification: A Comparison of Approaches* (Sheffield: University of Sheffield, Postgraduate School of Librarianship and Information Science, 1976); Jessie B. Watkins, *Selected Bibliography on Maps in Libraries: Acquisition, Classification, Cataloging, Storage, Uses* (Syracuse: Syracuse University Libraries, 1967).

14. Here the literature is surprisingly sparse. See Richardson, *Classification: Theoretical and Practical*; Sayers, *An Introduction to Library Classification*; in geography, Merrett, *Map Cataloguing and Classification*; and for an account relating classification to the practice of science, Anne Brearley Piternick, "Traditional Interpretations of 'Authorship' and 'Responsibil-

ity' in the Description of Scientific and Technical Documents," *Cataloguing and Classification Quarterly* 5 (1985): 17–33.

15. Michel Foucault, "What Is an Author?" in Josué Harari, ed., *Textual Strategies: Perspectives in Post-Structuralist Criticism* (Ithaca, N.Y.: Cornell University Press, 1979), pp. 141–60; Mark Rose, *Authors and Owners: The Invention of Copyright* (Cambridge: Harvard University Press, 1993); Martha Woodmansee, "The Genius and the Copyright: Economic and Legal Conditions of the Emergence of the 'Author,'" *Eighteenth Century Studies* 17 (1984): 425–48.

16. Elizabeth Eisenstein, *The Printing Press as an Agent of Change: Communications and Cultural Transformations in Early Modern Europe* (Cambridge: Cambridge University Press, 1979); David Woodward, ed., *Five Centuries of Map Printing* (Chicago: University of Chicago Press, 1975).

17. Walter J. Ong, "System, Space, and Renaissance Symbolism," *Study on Renaissance Symbolism* 8 (1956): 229.

18. Walter Benjamin, "The Work of Art in the Age of Mechanical Reproduction," in *Illuminations*, ed. Hannah Arendt (New York: Schocken, 1969), pp. 217–51.

19. Arthur Robinson, "Mapmaking and Map Printing: The Evolution of a Working Relationship," in Woodward, *Five Centuries of Map Printing*, p. 14.

20. Rose, *Authors and Owners*, pp. 4–5. See also Lyman Ray Patterson, *Copyright in Historical Perspective* (Nashville: Vanderbilt University Press, 1968); Harry Ransom, *The First Copyright Statute: An Essay on the Act for the Encouragement of Learning, 1710* (Austin: University of Texas Press, 1956); and Peter Jaszi, "On the Author Effect: Contemporary Copyright and Collective Creativity," *Cardozo Arts and Entertainment Journal* 10 (1992): 293–320.

21. Rose, *Authors and Owners*, p. 56.

22. Ibid., p. 65.

23. John Locke, "Second Treatise of Government," in *Two Treatises of Government*, ed. Thomas I. Cook (New York: Hafner, 1947).

24. C. B. MacPherson, *The Political Theory of Possessive Individualism: Hobbes to Locke* (Oxford: Oxford University Press, 1962); John R. Wikse, *About Possession: The Self as Private Property* (University Park: Pennsylvania State University Press, 1977).

25. John Locke, *An Essay Concerning Human Understanding* (London: Routledge, n.d.); David Hume, *Enquiries Concerning Human Understanding and Concerning the Principles of Morals*, 3d ed. (Oxford: Oxford University Press, 1975).

26. See, for example, Terry Eagleton, "A Small History of Rhetoric," in *Walter Benjamin, or, Towards a Revolutionary Criticism* (London: NLB-Verso, 1981), pp. 101–13; Brian Vickers, *In Defence of Rhetoric* (Oxford: Clarendon Press, 1988); and Charles W. Kneupper, ed., *Visions of Rhetoric: History, Theory, and Criticism* (Arlington, Tex.: Rhetoric Society of America, 1987).

27. Woodmansee, "The Genius and the Copyright," p. 429.

28. Johann Gottlieb Fichte, "Proof of the Illegality of Reprinting," pp. 227–28, quoted in Woodmansee, "The Genius and the Copyright," p. 445.

29. Lawrence C. Becker, "The Labor Theory of Property Acquisition," *Journal of Philosophy* 73 (1976): 653–63; David Ellerman, "On the Labor Theory of Property," *Philosophical Forum* 16 (1985): 293–326; Edwin C. Hettinger, "Justifying Intellectual Property," *Philosophy and Public Affairs* 18 (1989): 3–52; Alfred C. Yen, "Restoring the Natural Law: Copyright as Labor and Possession," *Ohio State Law Journal* 51 (1990): 517–59. I shall return to this view in a later chapter.

30. Thomas C. Grey, "The Disintegration of Property," in J. Roland Pennock and John William Chapman, eds., *Property* (New York: New York University Press, 1980), pp. 69–85.

31. Henry Barschall, "The Cost-Effectiveness of Physics Journals," *Physics Today* 41 (1988): 56–59. See also William Jaco, "Editorial," *Notices of the American Mathematical Society* 37 (1990): 2f; Tom Gaughan, "Editorial: A Splendid Wake for a Corporate Suicide," *American*

Libraries (1990): 172; and "Survey of American Research Journals," *Notices of the American Mathematical Society* 36 (1989): 1193–98.

32. Judith Axler Turner, "Library Survey from Journal Publisher Sent under Unsuspecting Foundation's Name," *Chronicle of Higher Education*, February 21, 1990: A6.

33. Gordon & Breach, "Paid Advertisement," *Notices of the American Mathematical Society* 37 (1990): 92–93.

34. Turner, "Library Survey from Journal Publisher Sent under Unsuspecting Foundation's Name," p. A6.

35. "Serials Pricing Controversy Escalates to Lawsuits," *UCLA Library News for the Faculty* 5 (1990): 4.

36. Martin Heidegger, "The Age of the World Picture," in *The Question Concerning Technology and Other Essays*, trans. William Lovitt (New York: Garland, 1977), pp. 115–54.

3. Formulating the New Common Sense

1. Trevor J. Barnes, "Homo Economicus, Physical Metaphors, and Universal Models in Economic Geography," *Canadian Geographer* 31 (1987): 299–308; Trevor J. Barnes and Michael R. Curry, "Post-Modernism in Economic Geography: Metaphor and the Construction of Alterity," *Environment and Planning D: Society and Space* 10 (1992): 57–68; Trevor J. Barnes and James S. Duncan, eds., *Writing Worlds: Discourse, Text, and Metaphor in the Representation of Landscape* (London: Routledge, 1992); Anne Buttimer, "Musing on Helicon: Root Metaphors and Geography," *Geoscience and Man* 24 (1984): 55–62; David N. Livingstone and Richard T. Harrison, "The Frontier: Metaphor, Myth, and Model," *Professional Geographer* 32 (1980): 127–32; Claude Raffestin, "Québec comme metaphore," *Cahiers de géographie du Québec* 25 (1981): 61–69.

2. Max Black, *Models and Metaphors* (Ithaca: Cornell University Press, 1954); Norwood Russell Hanson, *Patterns of Discovery* (Cambridge: Cambridge University Press, 1958). See more recently but in the same vein, Mary Gerhart and Allan Melvin Russell, *Metaphoric Process: The Creation of Scientific and Religious Understanding* (Fort Worth: Texas Christian University Press, 1984).

3. Here see Black, and especially Douglas Berggren, "The Use and Abuse of Metaphor," *Review of Metaphysics* 16 (1962–63): 237–58, 450–72.

4. Mary B. Hesse, "The Explanatory Function of Metaphor," in *Revolutions and Reconstructions in the Philosophy of Science* (Bloomington: Indiana University Press, 1980), pp. 111–24; Geoffrey N. Cantor, "Weighing Light: The Role of Metaphor in Eighteenth Century Optical Discourse," in Andrew E. Benjamin, Geoffrey N. Cantor, and John R. R. Christie, eds., *The Figural and the Literal: Problems of Language in the History of Science and Philosophy, 1630–1800* (Manchester: Manchester University Press, 1987), pp. 124–46.

5. Stephen Pepper, "The Root Metaphor Theory of Metaphysics," in Warren Shibles, ed., *Essays on Metaphor* (Whitewater, Wis.: Language Press, 1972), pp. 15–26; Stephen Pepper, *World Hypothesis: A Study in Evidence* (Berkeley: University of California Press, 1942); George Lakoff and Mark Johnson, *Metaphors We Live By* (Chicago: University of Chicago Press, 1980).

6. M. J. Egenhofer, "Why Not SQL," *International Journal of Geographical Information Systems* 6 (1992): 71–85; K. E. Foote, review of *Landscapes of the Mind: Worlds of Sense and Metaphor*, by J. D. Porteous, *Annals of the Association of American Geographers* 82 (1992): 177–80; R. J. Horvath, "National-Development Paths 1965–1987: Measuring a Metaphor," *Environment and Planning A* 26 (1994): 285–305.

7. Gunnar Olsson, "Braids of Justification," unpublished MS, 1989.

8. Friedrich Nietzsche, "On Truth and Lie in an Extra-Moral Sense," in Walter Kaufmann, ed., *The Portable Nietzsche* (New York: Viking Press, 1954), pp. 46–47.

9. Olsson, "Braids of Justification."

10. Charles Bazerman, *Shaping Written Knowledge: The Genre and Activity of the Experimental Article in Science* (Madison: University of Wisconsin Press, 1988); Alan G. Gross, *The Rhetoric of Science* (Cambridge: Harvard University Press, 1990); Lawrence J. Prelli, *A Rhetoric of Science: Inventing Scientific Discourse* (Columbia: University of South Carolina Press, 1989).

11. Joseph Gusfield, "The Literary Rhetoric of Science: Comedy and Pathos in Drinking Driver Research," *American Sociological Review* 41 (1976): 16–34; see also the more extended version in Joseph Gusfield, *The Culture of Public Problems: Drinking-Driving and the Symbolic Order* (Chicago: University of Chicago Press, 1981).

12. Prelli, *A Rhetoric of Science*, p. 8.

13. Gross, *The Rhetoric of Science*, p. 20.

14. Alan G. Gross, "An English Professor Looks at the Scientific Article," review of *Shaping Written Knowledge: The Genre and Activity of the Experimental Article in Science*, by Charles Bazerman, *Studies in the History and Philosophy of Science* 21, no. 2 (1990): 347.

15. John Bender and David E. Wellbery, "Rhetoricality: On the Modernist Turn of Rhetoric," in *The Ends of Rhetoric: History, Theory, Practice* (Stanford: Stanford University Press, 1990), p. 25.

16. Julia Kristeva, "Word, Dialogue, and Novel," in *The Kristeva Reader*, ed. Toril Moi (New York: Columbia University Press, 1986), p. 37.

17. Olsson, "Braids of Justification," p. 1.

18. Richard Rorty, *Philosophy and the Mirror of Nature* (Princeton: Princeton University Press, 1979); Richard Rorty, *The Consequences of Pragmatism (Essays: 1972–80)* (Minneapolis: University of Minnesota Press, 1982).

19. Michael J. Dear, "The Postmodern Challenge: Reconstructing Human Geography," *Transactions, Institute of British Geographers* NS 13 (1988): 266.

20. Philip Cooke, "Local Capacity and Global Restructuring: Some Preliminary Results from the CURS Research Programme" (paper presented at the sixth Urban Change and Conflict Conference, University of Kent, Canterbury, Sept. 20–23, 1987), p. 5.

21. Richard Symanski, "Why We Should Fear Postmodernists," *Annals of the Association of American Geographers* 84 (1994): 301–3.

22. Doreen Massey, "Flexible Sexism," *Environment and Planning D: Society and Space* 9 (1991): 31–57. See also Liz Bondi, "Feminism, Postmodernism, and Geography: Space for Women?" *Antipode* 22 (1990): 157–65; and Rosalind Deutsche, "Boys Town," *Environment and Planning D: Society and Space* 9 (1991): 5–30.

23. Richard Rorty, "Science as Solidarity," in John S. Nelson, Allan Megill, and Donald McCloskey, eds., *The Rhetoric of the Human Sciences: Language and Argument in Scholarship and Public Affairs* (Madison: University of Wisconsin Press, 1987), pp. 45–46; see also Rorty, *The Consequences of Pragmatism*; and Rorty, *Philosophy and the Mirror of Nature.*

24. Clifford Geertz, *Local Knowledge: Further Essays in Interpretive Anthropology* (New York: Basic Books, 1983).

25. David Bloor, "The Strengths of the Strong Programme," *Philosophy of the Social Sciences* 11 (1981): 199–213; David Bloor, *Wittgenstein: A Social Theory of Knowledge* (New York: Columbia University Press, 1983).

26. Michael J. Dear, "The Postmodern Challenge," pp. 265–66.

27. A useful introduction to the varieties of relativism is Martin Hollis and Steven Lukes, introduction to *Rationality and Relativism* (Cambridge: MIT Press, 1982), pp. 1–20; see also the essays in the volume, as well as Bryan Wilson, ed., *Rationality* (Oxford: Basil Blackwell, 1970), to which it is a sequel.

28. Richard Symanski, "The Manipulation of Ordinary Language," *Annals of the Association of American Geographers* 66 (1976): 605–14.

29. Gunnar Olsson, *Birds in Egg* (Ann Arbor: Department of Geography, University of Michigan, 1975).

30. Gunnar Olsson, "On Yearning for Home: An Epistemological View of Ontological Transformations," in Douglas C. D. Pocock, ed., *Humanistic Geography and Literature: Essays on the Experience of Place* (London: Croom Helm, 1981); Gunnar Olsson, "Toward a Sermon of Modernity," in Mark Billinge, Derek Gregory, and Ron Martin, eds., *Recollections of a Revolution: Geography as a Spatial Science* (New York: St. Martin's Press, 1983), pp. 73–85.

31. Denis Cosgrove and Mona Domosh, "Author and Authority: Writing the New Cultural Geography," in James S. Duncan and David Ley, eds., *Place/Culture/Representation* (London: Routledge, 1993), pp. 25–38.

32. Susan Hanson, "Soaring," *Professional Geographer* 40 (1988): 4–7; R. Cole Harris, "The Historical Mind and the Practice of Geography," in David Ley and Marwyn Samuels, eds., *Humanistic Geography: Prospects and Problems* (Chicago: Maaroufa Press, 1978), pp. 123–37.

33. Cosgrove and Domosh, "Author and Authority," p. 36.

34. Ibid., p. 29.

35. Ibid., p. 31.

36. Stephen A. Tyler, "Post-Modern Ethnography: From the Document of the Occult to the Occult Document," in James Clifford and George Marcus, eds., *Writing Culture* (Berkeley: University of California Press, 1986), pp. 122–40; and Stephen A. Tyler, *The Unspeakable: Discourse, Dialogue, and Rhetoric in the Postmodern World* (Madison: University of Wisconsin Press, 1987).

37. Trevor J. Barnes, "Homo Economicus, Physical Metaphors, and Universal Models in Economic Geography," *Canadian Geographer* 31 (1987): 299–308; Trevor J. Barnes, "Metaphors and Conversations in Economic Geography: Richard Rorty and the Gravity Model," *Geografiska Annaler B* 73 (1991): 111–20; Trevor J. Barnes, "Reading the Texts of Theoretical Economic Geography: The Role of Physical and Biological Metaphors," in Trevor J. Barnes and James S. Duncan, eds., *Writing Worlds*, pp. 118–35; and Trevor J. Barnes, "Rhetoric, Metaphor, and Mathematical Modelling," *Environment and Planning A* 26 (1994): 1021–40.

38. Richard Rorty, "Hesse and Davidson on Metaphor," *Proceedings of the Aristotelian Society Supplementary Volume* 61 (1987): 283–96; Richard Rorty, *Contingency, Irony, and Solidarity* (Cambridge: Cambridge University Press, 1989); Donald Davidson, "What Metaphors Mean," in Sheldon Sacks, ed., *On Metaphor* (Chicago: University of Chicago Press, 1979), pp. 29–46.

39. Barnes, "Metaphors and Conversations in Economic Geography," p. 114.

40. Barnes, "Reading the Texts of Theoretical Economic Geography."

41. Bloor, *Wittgenstein: A Social Theory of Knowledge*; Ludwig Wittgenstein, *Remarks on the Foundations of Mathematics*, rev. ed., ed. Georg Henrik Von Wright, Rush Rhees, and G. E. M. Anscombe (Cambridge: MIT Press, 1983).

42. Barnes, "Reading the Texts of Theoretical Economic Geography," p. 1033.

43. J. B. Harley, "Cartography, Ethics, and Social Theory," *Cartographica* 27 (1990): 1–23; J. B. Harley, "Deconstructing the Map," *Cartographica* 26 (1989): 1–20; J. B. Harley, "Historical Geography and the Cartographic Illusion," *Journal of Historical Geography* 15 (1989): 80–91; J. B. Harley, "Maps, Knowledge, and Power," in Denis Cosgrove and Stephen Daniels, eds., *The Iconography of Landscape: Essays on the Symbolic Representation, Design, and Use of Past Environments* (Cambridge: Cambridge University Press, 1988), pp. 277–312; J. B. Harley, "Silences and Secrecy: The Hidden Agenda of Cartography in Early Modern Europe," *Imago Mundi* 40 (1988): 57–76.

44. Barbara Belyea, "Images of Power: Derrida/Foucault/Harley," *Cartographica* 29 (1992): 1–9.

45. Harley, "Deconstructing the Map," p. 5.

46. Ibid., pp. 5–6.

47. Ibid., p. 11.

48. Harley, "Silences and Secrecy," p. 58.

49. Ibid., p. 66.

50. Denis Wood, *The Power of Maps* (New York: Guilford, 1992); John Pickles, *Ground Truth: The Social Implications of Geographic Information Systems* (New York: Guilford, 1995).

51. For example, there is no discussion of the written work in founding works such as Janice Monk, "Integrating Women into the Geography Curriculum," *Journal of Geography* 82 (1983): 271–73; Linda McDowell and Sophia Bowlby, "Teaching Feminist Geography," *Journal of Geography in Higher Education* 7 (1983): 97–108; Linda Peake, "Teaching Feminist Geography: Another Perspective," *Journal of Geography in Higher Education* 9 (1985): 186–90; and Women and Geography Study Group, *Geography and Gender: An Introduction to Feminist Geography* (London: Hutchinson Press, 1984).

52. Mona Domosh, "Toward a Feminist Historiography of Geography," *Transactions, Institute of British Geographers* NS 16 (1991): 98.

53. Massey, "Flexible Sexism," pp. 31–57; Deutsche, "Boys Town," pp. 5–30; and Steve Pile and Gillian Rose, "All or Nothing? Politics and Critique in the Modernism-Postmodernism Debate," *Environment and Planning D: Society and Space* 10 (1992): 123–36.

54. Pile and Rose, "All or Nothing?" pp. 126–27; the quotation is from Gayatri Spivak, "Can the Subaltern Speak?" in C. Nelson and L. Grossberg, eds., *Marxism and the Interpretation of Culture* (Macmillan: Basingstoke, 1988), p. 280.

55. Deutsche, "Boys Town," p. 8.

56. Pile and Rose, "All or Nothing?" p. 129.

57. Massey, "Flexible Sexism," p. 36. The works quoted here are Edward W. Soja, *Postmodern Geographies: The Reassertion of Space in Critical Social Theory* (London: Verso, 1989); and Fredric Jameson, "Marxism and Postmodernism," *New Left Review 136* (1989): 31–45.

58. Vincent Crapanzano, *Tuhami: Portrait of a Moroccan* (Chicago: University of Chicago Press, 1980); Kevin Dwyer, *Moroccan Dialogues* (Baltimore: Johns Hopkins University Press, 1982); Marjorie Shostak, *Nisa: The Life and Words of a !Kung Woman* (Cambridge: Harvard University Press, 1981).

59. Frances E. Mascia-Lees, Patricia Sharpe, and Colleen Ballerino Cohen, "The Postmodernist Turn in Anthropology: Cautions from a Feminist Perspective," *Signs: Journal of Women in Culture and Society* 15 (1989): 7–8.

60. Susan Christopherson, "On Being Outside 'the Project,'" *Antipode* 21 (1989): 84.

61. Linda McDowell, "Coming in from the Dark: Feminist Research in Geography," in John Eyles, ed., *Research in Human Geography: Introductions and Investigations* (Oxford: Basil Blackwell, 1988), p. 159.

62. Jane Flax, "Postmodernism and Gender Relations in Feminist Theory," in Linda J. Nicholson, ed., *Feminism/Postmodernism* (New York: Routledge, 1990) pp. 39–62.

63. Sandra Harding, "Feminism, Science, and the Anti-Enlightenment Critique," in Nicholson, ed., *Feminism/Postmodernism*, pp. 83–106.

64. Harding, "Feminism, Science, and the Anti-Enlightenment Critique," p. 95.

65. Ibid., p. 97.

66. Ibid., p. 98.

67. Ibid.

68. Flax, "Postmodernism and Gender Relations in Feminist Theory," pp. 56–57.

4. Beyond the New Common Sense: Toward a Geography of the Work in the World

1. Preston Everett James and Geoffrey J. Martin, *All Possible Worlds: A History of Geographical Ideas,* 2d ed. (New York: Wiley, 1981) provides a useful chronology of ancient geography. Edward Herbert Bunbury, *A History of Ancient Geography among the Greeks and Romans,* 2d. ed. (New York: Dover, 1959) is dated, and in a way more useful now for what it tells us about his own era, but is still useful. From within the history of science, George Sarton, "Geography and Chronology in the Third Century: Eratosthenes of Cyrene," in *A History of Science* (Cambridge: Harvard University Press, 1959), pp. 99–116, is the standard work. A more recent and very useful summary is David C. Lindberg, *The Beginnings of Western Science: The European Scientific Tradition in Philosophical, Religious, and Institutional Context, 600 B.C. to A.D. 1450* (Chicago: University of Chicago Press, 1992). G. E. R. Lloyd's works, *Early Greek Science: Thales to Aristotle* (London: Chatto and Windus, 1970); *Greek Science after Aristotle* (New York: W. W. Norton, 1973); *Magic, Reason, and Experience* (Cambridge: Cambridge University Press, 1979); *The Revolutions of Wisdom: Studies in the Claims and Practice of Ancient Greek Science* (Berkeley: University of California Press, 1987) are always interesting and provocative. Finally, two useful works more explicitly concerned with physics and astronomy are Michael Crowe, *Theories of the World from Antiquity to the Copernican Revolution* (New York: Dover, 1990); and Olaf Pederson, *Early Physics and Astronomy,* rev. ed. (Cambridge: Cambridge University Press, 1993).

2. Sarton, *A History of Science,* p. 102.

3. Fred Lukermann, "The Concept of Location in Classical Geography," *Annals of the Association of American Geographers* 51 (1961): 194–210.

4. The essential works here are the *Physics* (Physica) and *On the Heavens* (De Caelo), both available in many editions. I use Aristotle, "De Caelo," in Richard McKeon, ed., *The Basic Works of Aristotle,* trans. J. L. Stocks (New York: Random House, 1941), pp. 398–469; and Aristotle, *Physics,* trans. Richard Hope (Lincoln: University of Nebraska Press, 1961). The literature on his scientific work is quite massive; see the bibliography in Pederson, *Early Physics and Astronomy.*

5. There is a substantial body of work on medieval conceptions of space; here Edward Grant stands out. He has published a long series of articles and books; see especially Edward Grant, *Much Ado about Nothing: Theories of Space and Vacuum from the Middle Ages to the Scientific Revolution* (New York: Cambridge University Press, 1981); see also Edward Grant, "Medieval and Seventeenth-Century Conceptions of an Infinite Void Space beyond the Cosmos," *Isis* 60 (1969): 39–60; "Motion in the Void and the Principle of Inertia in the Middle Ages," *Isis* 55 (1964): 265–92; and "Place and Space in Medieval Physical Thought," in Peter K. Machamer and Robert G. Turnbull, eds., *Motion and Time, Space and Matter: Interrelations in the History of Philosophy and Science* (Columbus: Ohio State University Press, 1976), pp. 137–67.

6. O. A. W. Dilke, *The Roman Land Surveyors: An Introduction to the Agrimensores* (New York: Barnes and Noble, 1971).

7. René Descartes, *Principles of Philosophy* (1644), in G. E. M. Anscombe and Peter Geach, ed. and trans., *Descartes' Philosophical Writings* (Indianapolis: Bobbs-Merrill, 1971), II, § 10.

8. Ibid., II, § 12.

9. Ibid., II, § 21.

10. Ibid., II, § 21–22.

11. Isaac Newton, *The Mathematical Principles of Natural Philosophy and His System of the World* (Berkeley: University of California Press, 1934).

12. Ibid., Scholium § 2.

13. Ibid., Scholium § 4.

14. Ibid., Scholium § 4.

15. Isaac Newton, "General Scholium," in *The Mathematical Principles of Natural Philosophy and His System of the World.* The quotation is taken from Isaac Newton, *Newton's Philosophy of Nature: Selections from His Writings,* trans. H. S. Thayer (New York: Hafner, 1953), pp. 43–44.

16. H. G. Alexander, ed., *The Leibniz-Clarke Correspondence, Together with Extracts from Newton's Principia and Opticks* (New York: Barnes and Noble Imports, 1956).

17. Ibid., Leibniz, fifth paper, § 47.

18. Ibid.

19. It is important here to note that although the arguments of Descartes, Newton, and Leibniz can be easily translated into contemporary secular and scientific terms, the work of each was in fact steeped in a concern about the relationship of these issues to religious belief. The Leibniz-Clarke correspondence, for example, begins with Leibniz noting that Newton had called space the sensorium of God; he replies that this is utterly preposterous. Similarly, Leibniz himself argues against the existence of a vacuum on the basis that God would, and in a sense could, not have created empty space; appealing to what A. O. Lovejoy has called the principle of plenitude, he argues that God must have created everything that could have been created, and that empty space was a denial of God's power.

Although this set of arguments now seems quaint, Leibniz does refer to a principle, of sufficient reason, that has been seen in a secular twentieth-century version to have general applicability. As codified in the seventeenth century, this principle holds that God must have had a sufficient reason for making things one way rather than another. At that time the consequence, it was argued by Leibniz, was that the idea of absolute and empty space left no reason for an object's being here rather than there, and one was thereby forced to adopt a relational view of space. A twentieth-century version of this, albeit one that has not been applied to the question of space, is the principle of the identity of indiscernibles. It holds that if we have no means of distinguishing between two objects, then there must in fact only be a single object. This view was fundamental to thinking, in the form of the verification principle, in early twentieth-century logical positivist philosophy of science.

20. Alexander, *The Leibniz-Clarke Correspondence,* Leibniz, third paper, § 4.

21. Kant had in fact written his inaugural dissertation on space. That work, titled "On the First Ground of the Distinction of Regions in Space," in *Kant's Inaugural Dissertation and Early Writings on Space,* trans. John Handyside (Chicago: Open Court, 1929), has been widely seen as the expression of an earlier and abandoned view of space; scholars have generally divided his work into two periods, with the 1781 publication of the *Critique of Pure Reason,* trans. Norman Kemp Smith (New York: St. Martin's Press, 1965), signaling a change of heart. More recent work, though, suggests that with respect to the issue of space, there may in fact be more continuity than has been believed; see Paul Guyer, *Kant and the Claims of Knowledge* (Cambridge: Cambridge University Press, 1987).

22. Kant, *Critique of Pure Reason,* A23/B38.

23. There is, of course, another way in which geographers have appealed to Kant. Drawing from his assertion that the world is first organized in terms of space and time and then in terms of concepts, they have suggested that geography and history are the "exceptional" studies of the world, in terms of space and time, and that other sciences, like physics, operate within the subsequent conceptual realm. We find this view widely in discussions that attempt to "define geography" conceptually. Unfortunately, these works (Richard Hartshorne's discussions of the discipline are an example) typically miss the real significance of his understanding of the nature of space, which is its relationship through neo-Kantianism to the development of the theory of culture.

24. Paul Vidal de la Blache, *Principles of Human Geography,* trans. M. T. Bingham (London: Constable, 1926); Paul Vidal de la Blache, "Les Genres de vie dans la géographie humaine," *Annales de Géographie* 20 (1911): 193–212,289–304; Carl O. Sauer, "Morphology of Landscape," in *Land and Life: A Selection from the Writings of Carl Ortwin Sauer,* ed. John Leighly (Berkeley: University of California Press, 1963), pp. 315–50; Yi-Fu Tuan, *Topophilia: A Study of Environmental Perception, Attitudes, and Values* (Englewood Cliffs, N.J.: Prentice-Hall, 1974); Yi-Fu Tuan, *Space and Place: The Perspective of Experience* (Minneapolis: University of Minnesota Press, 1977).

25. Sauer, "Morphology of Landscape," pp. 315–50.

26. Paul Carter, *The Road to Botany Bay* (Chicago: University of Chicago Press, 1987).

27. Terry Alford, "The West as a Desert in American Thought prior to Long's 1819–20 Expedition," *Journal of the West* 8 (1969): 1–11; Martyn J. Bowden, "The Great American Desert and the American Frontier, 1800–1882: Popular Images of the Plains," in Tamara K. Hareven, ed., *Anonymous Americans: Explorations in Nineteenth-Century Social History* (Englewood Cliffs, N.J.: Prentice-Hall, 1971), pp. 48–79; Merlin P. Lawson and Charles W. Stockton, "Desert Myth and Climatic Reality," *Annals of the Association of American Geographers* 71 (1981): 527–35; G. Malcolm Lewis, "Three Centuries of Desert Concepts in the Cis-Rocky Mountain West," *Journal of the West* 4 (1965): 457–68.

28. Arthur S. Keller, Oliver J. Lissitzyn, and Frederick J. Mann, *Creation of Rights of Sovereignty through Symbolic Acts* (New York: Columbia University Press, 1938), p. 40.

29. Ibid., p. 41.

30. Yi-Fu Tuan, "Rootedness versus Sense of Place," *Landscape* 24 (1980): 3–8.

31. Eduard Jan Dijksterhuis, *The Mechanization of the World Picture,* trans. C. Dikshoorn (New York: Oxford University Press, 1961); see also Alexandre Koyré, *From the Closed World to the Infinite Universe* (Baltimore: Johns Hopkins University Press, 1957).

32. See Frederick W. Taylor, "Principles of Scientific Management," in *Scientific Management* (New York: Harper and Row, 1947); and Frank B. Gilbreth, *Motion Study: A Method for Increasing the Efficiency of the Workman* (New York: Van Nostrand, 1911). For an account of the relation of Fordism, Taylorism, and American culture, see Thomas Parke Hughes, *American Genesis: A Century of Invention and Technological Enthusiasm* (Harmondsworth: Penguin, 1989).

33. David F. Noble, *Forces of Production: A Social History of Industrial Automation* (New York: Oxford University Press, 1984).

34. Shoshana Zuboff, *In the Age of the Smart Machine: The Future of Work and Power* (New York: Basic Books, 1988).

35. Ian Burton and Robert W. Kates, "The Perception of Natural Hazards in Resource Management," *Natural Resources Journal* 3 (1964): 412–41; Robert W. Kates, *Hazard and Choice Perception on Flood Plain Management,* Research Paper No. 78 (Chicago: University of Chicago, Department of Geography, 1962); Thomas F. Saarinen, *Perception of the Drought Hazard on the Great Plains* (Chicago: University of Chicago, Department of Geography, 1966).

36. For example, Trevor J. Barnes and James S. Duncan, eds., *Writing Worlds: Discourse, Text, and Metaphor in the Representation of Landscape* (London: Routledge, 1992); Denis Cosgrove, *Social Formation and Symbolic Landscape* (Totowa, N.J.: Barnes and Noble, 1985); J. Nicholas Entrikin, *The Betweenness of Place* (Basingstoke: Macmillan, 1991); Yi-Fu Tuan, *Segmented Worlds and Self: Group Life and Individual Consciousness* (Minneapolis: University of Minnesota Press, 1982).

37. Richard Rorty, *Philosophy and the Mirror of Nature* (Princeton: Princeton University Press, 1979).

38. Peter Winch, *The Idea of a Social Science and Its Relation to Philosophy,* 2d ed. (New York: Humanities Press, 1990); Michel de Certeau, *The Practice of Everyday Life,* trans. Steven Rendell (Berkeley: University of California Press, 1984); Anthony Giddens, *Central Problems in Social Theory* (Berkeley: University of California Press, 1979); Pierre Bourdieu, *The Logic of Practice,* trans. Richard Nice (Cambridge: Polity, 1990).

39. David Bloor, *Wittgenstein: A Social Theory of Knowledge* (New York: Columbia University Press, 1983); H. M. Collins, *Artificial Experts: Social Knowledge and Intelligent Machines* (Cambridge: MIT Press, 1990); Andrew Pickering, "Big Science as a Form of Life," in M. De Maria, M. Grilli, and Fabio Sebastiani, eds., *The Restructuring of the Physical Sciences in Europe and the United States, 1945–1960* (Singapore: World Scientific Publishing, 1989), pp. 42–54; Andrew Pickering, ed., *Science as Practice and Culture* (Chicago: University of Chicago Press, 1992).

40. I should say that what I offer here ought not to be seen as an attempt to engage into an inquiry into "what Wittgenstein really meant." His work is notoriously difficult and inevitably requires explication, but my own explication can better be seen as an attempt to develop a point of view that is my own.

In keeping with common usage, I shall refer to Wittgenstein's works in the following way: *PI* = *Philosophical Investigations*; *Z* = *Zettel*; *LCAPRB* = *Lectures and Conversations on Aesthetics, Psychology, and Religious Belief*; and *T* = *Tractatus Logico-Philosophicus.* In the case of *PI,* references are preceded by a part number; references to part I (and to all of *Z* and *T*) are to sections; references to part II of *PI* and to *LCAPRB* are to page numbers.

41. Saul Kripke, *Wittgenstein on Rules and Private Language* (Cambridge: Harvard University Press, 1982), p. 97.

42. Michael Lynch, "Extending Wittgenstein: The Pivotal Move from Epistemology to the Sociology of Science," in Andrew Pickering, ed., *Science as Practice and Culture,* p. 223; summarizing David Bloor, *Wittgenstein: A Social Theory of Knowledge.*

43. Ernst Mach, *The Science of Mechanics: A Critical and Historical Account of its Development,* 6th ed., with revisions through the 9th German ed., trans. Thomas J. McCormack (La Salle, Ill.: Open Court, 1960). More recent analysts have made it fairly certain that Wittgenstein's work actually shared very little with that of the logical positivists, but that was not known at the time. See, for example, Allan Janik and Stephen Toulmin, *Wittgenstein's Vienna* (New York: Simon and Schuster, 1973); Rudolf Haller, *Questions on Wittgenstein* (Lincoln: University of Nebraska Press, 1988).

44. Carl G. Hempel, "The Function of General Laws in History," in *Aspects of Scientific Explanation and Other Essays in the Philosophy of Science* (New York: Free Press, 1965), pp. 231–44.

45. Norman Malcolm, *Ludwig Wittgenstein: A Memoir* (London: Oxford University Press, 1966), p. 9.

46. K. T. Fann, *Wittgenstein's Conception of Philosophy* (Berkeley: University of California Press, 1971), p. 43.

47. Here see, for example, Herbert Fingarette, *Heavy Drinking: The Myth of Alcoholism as a Disease* (Berkeley: University of California Press, 1988) for a compelling case study of the way in which this is true.

48. Richard Rorty, *The Consequences of Pragmatism (Essays: 1972–80)* (Minneapolis: University of Minnesota Press, 1982); Rorty, *Philosophy and the Mirror of Nature.*

49. MS 219, p. 11, quoted in Anthony Kenny, "Wittgenstein on the Nature of Philosophy," in Brian F. McGuinness, ed., *Wittgenstein and His Times* (Chicago: University of Chicago Press, 1982), p. 13.

50. Kenny, "Wittgenstein on the Nature of Philosophy," p. 13.

51. Ibid., p. 25.

5. Authorship and the Construction of Authority

1. Michel Foucault, "What Is an Author?" in Josué Harari, ed., *Textual Strategies: Perspectives in Post-Structuralist Criticism* (Ithaca, N.Y.: Cornell University Press, 1979), pp. 141–60. Subsequent references will be given in the text in parentheses.

2. Peter Gould, *The Geographer at Work* (London: Routledge and Kegan Paul, 1985); Peter Haggett, *The Geographer's Art* (Oxford: Blackwell, 1990); Allan R. Pred, "The Academic Past through a Time-Geographic Looking Glass," *Annals of the Association of American Geographers* 69 (1979): 175–80.

3. Maeve O'Connor, "Style Manuals and Guides for Authors and Editors: Prescriptive or Descriptive?" in Miriam Balaban, ed., *Scientific Information Transfer: The Editor's Role. Proceedings of the First International Conference of Scientific Editors* (Dordrecht: Reidel, 1978), pp. 287–90.

4. Donald W. Meinig, "Geography as an Art," *Transactions, Institute of British Geographers* NS 8 (1983): 314–28.

5. Charles Bazerman, "Codifying the Social Scientific Style: The APA *Publication Manual* as a Behaviorist Rhetoric," in John S. Nelson, Allan Megill, and Donald McCloskey, eds., *The Rhetoric of the Human Sciences: Language and Argument in Scholarship and Public Affairs* (Madison: University of Wisconsin Press, 1987), pp. 125-44; Charles Bazerman, *Shaping Written Knowledge: The Genre and Activity of the Experimental Article in Science* (Madison: University of Wisconsin Press, 1988); Ted Cohen, "Metaphor and the Cultivation of Intimacy," in Sheldon Sacks, ed., *On Metaphor* (Chicago: University of Chicago Press, 1979), pp. 1–10.

6. Michel Foucault, *Discipline and Punish: The Birth of the Prison,* trans. Alan Sheridan (New York: Vintage Books, 1977); Michael J. Mahoney, "Bias, Controversy, and Abuse in the Study of the Scientific Publication System," *Science, Technology, and Human Values* 15 (1990): 50–55.

7. Asa Kashar, "Style! Why Bother?" in Miriam Balaban, ed., *Scientific Information Transfer,* pp. 299–301.

8. Anders Martinsson, Basil Walby, Ruth Weiss, Asa Kashar, Knut Faegri, Klaus Helbig, Philip L. Altman, and David Shephard, "Discussion of Altman and Kashar," in Miriam Balaban, ed., *Scientific information Transfer,* pp. 303–5.

9. Eugene Garfield, "'Science Citation Index'—A New Dimension in Indexing," *Science* 144 (1964): 649–54.

10. Editorial, "Information for Authors," *Annals of the Association of American Geographers* 72 (1982): ii–iii; see also Derek Gregory, editorial, *Environment and Planning D: Society and Space* (1990) 8: 1–6, which makes some of the same points that I make here.

11. Robert A. Day, *How to Write and Publish a Scientific Paper,* 2d ed. (Philadelphia: ISI Press, 1983), p. 5.

12. Martin Kenzer, ed., *On Becoming a Professional Geographer* (Columbus: Merrill, 1989).

13. Stanley D. Brunn, "Personal and General Publishing Policies of Geographers," *Terra* 99 (1987): 155–65; David R. Butler, "Conducting Research and Writing an Article in Physical Geography," in Kenzer, ed., *On Becoming a Professional Geographer,* p. 90.

14. Butler, "Conducting Research and Writing an Article in Physical Geography," p. 88.

15. L. S. Bourne, "On Writing and Publishing in Human Geography: Some Personal Reflections," in Kenzer, ed., *On Becoming a Professional Geographer,* pp. 100–112; Butler, "Conducting Research and Writing an Article in Physical Geography," pp. 88–99; Bonnie Loyd, "Not-So-Mysterious Secrets of Publishing Journal Articles," in Kenzer, ed., *On Becoming a Professional Geographer,* pp. 148–59.

16. See, for example, Daryl E. Chubin and Soumyo D. Moitra, "Content Analysis of References Adjunct or Alternative to Citation Counting?" *Social Studies of Science* 5 (1975): 423–41;

David Edge, "Quantitative Measures of Communication in Science: A Critical Review," *History of Science* 17 (1979): 102–34; Eugene Garfield, "Citation Analysis as a Tool in Journal Evaluation," *Science* 178 (1972): 471–79; Eugene Garfield, "Is Citation Analysis a Legitimate Evaluation Tool?" *Scientometrics* 1 (1979): 359–75; Garfield, " 'Science Citation Index,' " pp. 649–54; Norman Kaplan, "The Norms of Citation Behaviour: Prolegomena to the Footnote," *American Documentation* 16 (1965): 179–84. In geography, see J. W. R. Whitehand, "Contributors to the Recent Development and Influence of Human Geography: What Citation Analysis Suggests," *Transactions, Institute of British Geographers,* NS 10 (1985): 222–34; B. L. Turner II, and William B. Meyer, "The Use of Citation Indices in Comparing Geography Programs: An Exploratory Study," *Professional Geographer* 37 (1985): 271–78.

17. Whitehand, "Contributors to the Recent Development and Influence of Human Geography"; Turner and Meyer, "The Use of Citation Indices in Comparing Geography Programs"; Neil Wrigley and Stephen Matthews, "Citation Classics and Citation Levels in Geography," *Area* 18 (1986): 185–94; Neil Wrigley and Stephen Matthews, "Citation Classics in Geography and the New Centurions: A Response to Haigh, Mead, and Whitehand," *Area* 18 (1986): 279–84.

18. Anthony C. Gatrell and Anthony Smith, "Networks of Relations among a Set of Geographical Journals," *Professional Geographer* 36 (1984): 300–306.

19. Derek J. de Solla Price, "Networks of Scientific Papers," *Science* 149 (1965): 510–15.

20. Kaplan, "The Norms of Citation Behaviour."

21. Warren O. Hagstrom, *The Scientific Community* (New York: Basic Books, 1965); Bruno Latour and Steve Woolgar, "The Cycle of Credibility," in Barry Barnes and David Edge, eds., *Science in Context: Readings in the Sociology of Science* (Cambridge: MIT Press, 1982), pp. 35–43.

22. Derek J. de Solla Price, *Little Science, Big Science* (New York: Columbia University Press, 1963), p. 65.

23. Price, *Little Science, Big Science,* p. 8.

24. Ibid., p. 9; see also David A. Kronick, *A History of Scientific and Technical Periodicals: The Origins and Development of the Scientific and Technological Press, 1665–1790* (New York: Scarecrow Press, 1962).

25. Barry S. Barnes and R. G. A. Dolby, "The Scientific Ethos: A Deviant Viewpoint," *Archives Européennes de sociologie* 11 (1970): 3–25.

26. Garfield, " 'Science Citation Index.' "

27. Ibid., p. 649.

28. Dun's Marketing Services, *Million Dollar Directory: America's Leading Private and Public Companies,* series, 1989 (Parsippany, N.J.: Dun's Marketing Services, 1989); *Ward's Business Directory of U.S. Private and Public Companies* (Detroit: Gale Research, 1983).

29. Quoted in H. Burr Steinbach, "The Quest for Certainty: *Science Citation Index,*" *Science* 145 (1964): 43.

30. Galileo Galilei, *Dialogue Concerning Two New Sciences,* trans. Henry Crew and Alfonse de Salvio (New York: Dover, 1954); Galileo Galilei, *Dialogue Concerning the Two Chief World Systems—Ptolemaic and Copernican,* 2d ed., trans. Stillman Drake (Berkeley: University of California Press, 1967).

31. G. W. F. Hegel, *Philosophy of Right,* trans. T. M. Knox (London: Oxford University Press, 1967).

32. Jane C. Ginsburg, "French Copyright Law: A Comparative Overview," *Journal of the Copyright Society of the USA* 4 (1989): 269–85; Justin Hughes, "The Philosophy of Intellectual Property," *Georgetown Law Journal* 77 (1988): 287–366; Arthur S. Katz, "The Doctrine of Moral Rights and American Copyright Law—A Proposal," *Southern California Law Review* 24 (1951): 375–427; Martin A. Roeder, "The Doctrine of Moral Rights: A Study in the Law of

Artists, Authors, and Creators," *Harvard Law Review* 53 (1940): 554–78; Raymond Sarraute, "Current Theory on the Moral Right of Authors and Artists under French Law," *American Journal of Comparative Law* 16 (1968): 465–86.

33. John Locke, "Second Treatise of Government," in *Two Treatises of Government,* ed. Thomas I. Cook (New York: Hafner, 1947), ch. 5, sec. 44. See also Lawrence C. Becker, "The Labor Theory of Property Acquisition," *Journal of Philosophy* 73 (1976); David Ellerman, "On the Labor Theory of Property," *Philosophical Forum* 16 (1985): 293–326.

34. David Harvey, *The Condition of Postmodernity: An Enquiry into the Origins of Cultural Change* (New York: Basil Blackwell, 1989); David Harvey, *Explanation in Geography* (London: Edward Arnold, 1969); David Harvey, *The Limits to Capital* (Oxford: Basil Blackwell, 1982); David Harvey, *Social Justice and the City* (Baltimore: Johns Hopkins University Press, 1973).

35. Harvey, *Explanation in Geography,* pp. v-vi.

36. Harvey, *Social Justice and the City,* pp. 9–10.

37. Harvey, *The Condition of Postmodernity: An Enquiry into the Origins of Cultural Change,* p. viii.

38. René Descartes, "Discourse on the Method," in *Descartes' Philosophical Writings,* ed. G. E. M. Anscombe and Peter Geach (Indianapolis: Bobbs-Merrill, 1971), p. 8.

6. The Work in the World

1. Fritz Machlup and Kenneth Leeson, *Information through the Printed Word: The Dissemination of Scholarly, Scientific, and Intellectual Knowledge* (New York: Praeger, 1978), p. 235.

2. William D. Garvey, Nan Lin, and Carnot E. Nelson, "Some Comparisons of Communication Activities in the Physical and Social Sciences," in Carnot E. Nelson and Donald K. Pollock, eds., *Communication among Scientists and Engineers* (Lexington, Mass.: D. C. Heath, 1970), pp. 61–84; Nan Lin, William D. Garvey, and Carnot E. Nelson, "A Study of the Communication Structure of Science," in Nelson and Pollock, eds., *Communication among Scientists and Engineers,* pp. 23–60.

3. J. W. R. Whitehand, "Contributors to the Recent Development and Influence of Human Geography: What Citation Analysis Suggests," *Transactions, Institute of British Geographers* NS 10 (1985): 222–34.

4. Ibid., p. 224.

5. J. W. R. Whitehand, "An Assessment of 'Progress,'" *Progress in Human Geography* 14 (1990): 12–23.

6. Gina Kolata, "In a Frenzy, Math Enters Age of Electronic Mail," *New York Times,* June 26, 1990, B, p. 5 f.

7. Anthony C. Gatrell and Anthony Smith, "Networks of Relations among a Set of Geographical Journals," *Professional Geographer* 36, no. 3 (1984): 300–6.

8. Eric S. Sheppard, editor, *Antipode,* personal communication with author, May 1, 1990.

9. Langdon Winner, *Autonomous Technology: Technics-Out-of-Control as a Theme in Political Thought* (Cambridge: MIT Press, 1977).

10. Stanley D. Brunn, "Editorial: Ethics in Word and Deed," *Annals of the Association of American Geographers* 79 (1989): iii-iv.

11. Gillian Page, Robert Campbell, and Jack Meadows, *Journal Publishing: Principles and Practice* (London: Butterworths, 1987); Maurice B. Visscher, "Copyright and Other Impediments to Scientific Communication," in Stacey B. Day, ed., *Communication of Scientific Information* (Basel: S. Karger, 1975), pp. 118–28.

12. Norman Kaplan, "The Norms of Citation Behaviour: Prolegomena to the Footnote," *American Documentation* 16 (1965): 181.

13. Michael J. Moravcsik and Poovanalingam Murugesan, "Some Results on the Function and Quality of Citations," *Social Studies of Science* 5 (1975): 86–92.

14. Frederick C. Thorne, "The Citation Index: Another Case of Spurious Validity," *Journal of Clinical Psychology* 33 (1977): 1157–61.

15. J. B. Bavelas, "The Social Psychology of Citations," *Canadian Psychological Review* 19 (1978): 158–63.

16. Arthur C. Danto, "Narrative Sentences," *History and Theory* 2 (1962): 146–79.

17. Whitehand, "An Assessment of 'Progress,'" p. 21.

18. Eugene Garfield, "Citation Analysis as a Tool in Journal Evaluation," *Science,* 178 (1972): 471–79; Eugene Garfield, "Is Citation Analysis a Legitimate Evaluation Tool?" *Scientometrics* 1 (1979): 359–75; Henry G. Small, "Cited Documents as Concept Symbols," *Social Studies of Science* 8 (1978): 327–40; Edward Nadel, "Multivariate Citation Analysis and the Changing Cognitive Organization in a Specialty of Physics," *Social Studies of Science* 10 (1980): 449–73.

19. David Lee and Arthur Evans, "American Geographers' Rankings of American Geography Journals," *Professional Geographer* 36 (1984): 292–300; David Lee and Arthur Evans, "Geographers' Rankings of Foreign Geography and Non-geography Journals," *Professional Geographer* 37 (1985): 396–402; B. L. Turner II and William B. Meyer, "The Use of Citation Indices in Comparing Geography Programs: An Exploratory Study," *Professional Geographer* 37 (1985): 271–78; Whitehand, "Contributors to the Recent Development and Influence of Human Geography"; Neil Wrigley and Stephen Matthews, "Citation Classics and Citation Levels in Geography," *Area* 18 (1986): 185–94; Neil Wrigley and Stephen Matthews, "Citation Classics in Geography and the New Centurions: A Response to Haigh, Mead, and Whitehand," *Area* 18 (1986): 279–84.

20. Richard D. Whitley, "The Establishment and Structure of the Sciences as Reputational Organizations," in Norbert Elias, Herminio Martins, and Richard Whitley, eds., *Scientific Establishments and Hierarchies* (Dordrecht: D. Reidel, 1982), pp. 313–57.

21. Harold Garfinkel, Michael Lynch, and Eric Livingston, "The Work of a Discovering Science Construed with Materials from the Optically Discovered Pulsar," *Philosophy of the Social Sciences* 11 (1981): 132.

22. John Q. Stewart and William Warntz, "Macrogeography and Social Science," *Geographical Review* 48 (1958): 167–84.

23. The works in question are Anne Buttimer, "Grasping the Dynamism of the Lifeworld," *Annals of the Association of American Geographers* 66 (1976): 277–92; Michael J. Dear, "The Postmodern Challenge: Reconstructing Human Geography," *Transactions, Institute of British Geographers* NS 13 (1988): 262–274; Leonard Guelke, "An Idealist Alternative in Human Geography," *Annals of the Association of American Geographers* 64 (1974): 193–202; Allan R. Pred, "The Choreography of Existence: Comments on Hägerstrand's Time-Geography and Its Usefulness," *Economic Geography* 53 (1977): 207–21; Edward Relph, *Place and Placelessness* (London: Pion, 1976).

24. General works include Bruno Latour and Steve Woolgar, *Laboratory Life: The Social Construction of Scientific Facts* (Beverly Hills: Sage Publications, 1979); John Law, "On the Methods of Long-Distance Control: Vessels, Navigation, and the Portuguese Route to India," in *Power, Action, and Belief: A New Sociology of Knowledge?* (London: Routledge, 1986), pp. 235–63; Michel Callon, "Some Elements of a Sociology of Translation: Domestication of the Scallops and the Fishermen of St. Brieuc Bay," in John Law, ed., *Power, Action, and Belief.* For works more specifically dealing with texts, see the articles in Michel Callon, John Law, and Arie Rip, eds., *Mapping the Dynamics of Science and Technology: Sociology of Science in the Real World* (Basingstoke: Macmillan, 1986); as well as Michel Callon and John Law, "On Interests and Their Transformation: Enrolment and Counter-Enrolment," *Social Studies of Science* 12 (1982): 615–25; and John Law and R. J. Williams, "Putting Facts Together: A Study of Scientific Persuasion," *Social Studies of Science* 12 (1982): 535–58.

25. Guelke, "An Idealist Alternative in Human Geography," p. 193. Subsequent page references will be given in the text in parentheses.

26. R. G. Collingwood, *The Idea of History* (London: Oxford University Press, 1956).

27. Buttimer, "Grasping the Dynamism of the Lifeworld," p. 277. Subsequent page references will be given in the text in parentheses.

28. Pred, "The Choreography of Existence: Comments on Hägerstrand's Time-Geography and Its Usefulness," p. 207.

29. Ibid., p. 209.

30. Dear, "The Postmodern Challenge: Reconstructing Human Geography," p. 262.

31. Ibid., p. 267.

32. Stewart and Warntz, "Macrogeography and Social Science," p. 167. Subsequent page references will be given in the text in parentheses.

33. I should note that the characterization of ethical issues strictly within the discourse of rights and responsibilities is itself limiting. This characterization, an outgrowth of a variety of processes, is part and parcel of the elision of the legal and the moral. I emphasize that my discussion needs to be seen as a recognition of the importance of that fact and not as support of it. Indeed, it strikes me that a range of recent works, from Alasdair MacIntyre, *After Virtue: A Study in Moral Theory,* 2d ed. (Notre Dame: University of Notre Dame Press, 1984) on the conservative side to feminist works such as Evelyn Fox Keller, *Reflections on Gender and Science* (New Haven: Yale University Press, 1985) and Joan Greenbaum, "The Head and the Heart: Using Gender Analysis to Study the Social Construction of Computer Systems," *Computers and Society* 20 (1990): 9–18, offer compelling reasons for believing that the characterization of relations within the workplace in terms of the head-body image and the concomitant discourse of rights and responsibilities paint only an emaciated picture of those ethical relationships.

34. Robert K. Merton, "The Normative Structure of Science," in *The Sociology of Science: Theoretical and Empirical Investigations,* ed. Norman W. Storer (Chicago: University of Chicago Press, 1973), pp. 267–78.

35. Brunn, "Editorial: Ethics in Word and Deed"; Michael R. Curry, "On the Possibility of Ethics in Geography: Writing, Citing, and the Construction of Intellectual Property," *Progress in Human Geography* 15 (1991): 125–47.

36. Karl Marx, *Capital* (New York: International Publishers, 1967); Frederick W. Taylor, "Principles of Scientific Management," in *Scientific Management* (New York: Harper and Row, 1947); Harry Braverman, *Labor and Monopoly Capital: The Degradation of Work in the Twentieth Century* (New York: Monthly Review Press, 1974).

37. Norbert Elias, *The Civilizing Process,* trans. Edmund Jephcott (New York: Urizen Books, 1982); Shoshana Zuboff, *In the Age of the Smart Machine: The Future of Work and Power* (New York: Basic Books, 1988).

38. Ruth Perry and Lisa Greber, "Women and Computers: An Introduction," *Signs: Journal of Women in Culture and Society* 16 (1990): 74–101; Angelika Volst and Ina Wagner, "Inequality in the Automated Office: The Impact of Computers on the Division of Labour," *International Sociology* 3 (1988): 129–54.

39. Michael C. Gemignani, "More on the Use of Computers by Professionals," *Rutgers Computer and Technology Law Journal* 13 (1987): 317–39; Alan H. Goldman, *The Moral Foundations of Professional Ethics* (Totowa, N.J.: Rowman and Littlefield, 1980); John Ladd, "The Quest for a Code of Professional Ethics: An Intellectual and Moral Confusion," in Deborah G. Johnson and John W. Snapper, eds., *Ethical Issues in the Use of Computers* (Belmont, Calif.: Wadsworth Publishing, 1985), pp. 8–13; Vivian Weil, "The Rise of Engineering Ethics," *Technology in Society* 6 (1984): 341–45.

40. Michael Gemignani, "Copyright Protection: Computer-Related Dependent Works," *Rutgers Computer and Technology Law Journal* 15 (1989): 383–410.

41. Alan Garnham, *Artificial Intelligence: An Introduction* (London: Routledge, 1987); T. J. Kim, L. L. Wiggins, and J. R. Wright, eds., *Expert Systems: Applications in Urban Planning* (New York: Springer-Verlag, 1990); A. Waterman, *A Guide to Expert Systems* (Reading, Mass.: Addison-Wesley, 1986).

42. Christopher R. Muse, "Patented Personality," *Santa Clara Computer and High-Technology Law Journal* 4 (1988): 285–305.

43. Commission of the European Communities, *Proposal for a Council Directive on the Legal Protection of Databases* (Brussels: European Community, 1992) COM(92) 24 final–SYN 393 (13 May 1992); Council of Europe, "Proposal for a Council Directive Harmonizing the Term of Protection of Copyright and Certain Related Rights," *IIC: International Review of Industrial Property and Copyright Law* 23, no. 6 (1992): 806–11; Jonathan Band and L. F. H. McDonald, "The Proposed EC Database Directive: The Reversal of FEIST v. Rural Telephone," *Computer Lawyer* 9 (1992): 19–21; X. R. Lopez, "Database Copyright Issues in the European GIS Community," *Government Information Quarterly* 10 (1993): 305–18; Phillip H. Miller, "Life after FEIST: Facts, the First Amendment, and the Copyright Status of Automated Databases," *Fordham Law Review* 60 (1991): 507–39; Michael Pattison, "The European Commission's Proposal on the Protection of Computer Databases," *European Intellectual Property Review* 14 (1992): 113–20.

44. Michael H. Agranoff, "Curb on Technology: Liability for Failure to Protect Computerized Data against Unauthorized Access," *Santa Clara Computer and High-Technology Law Journal* 5 (1989): 263–320; T. Soloman, "Personal Privacy and the '1984' Syndrome," *New England Law Review* 7 (1985): 753, 760–71; Commission of the European Communities, *Proposal for a Council Directive Concerning the Protection of Individuals in Relation to the Processing of Personal Data* (Brussels: European Community, 1992) COM(90) 314 final–SYN 287 (90/C 277/03); Council of the European Communities, *Proposal for a Council Directive Concerning the Protection of Individuals in Relation to the Processing of Personal Data* (Brussels: European Community, 1990) (COM(90) 314 final-SYN 287) (11/5/90).

45. Rainer Born, ed., *Artificial Intelligence: The Case Against* (London: Routledge, 1987); Hubert L. Dreyfus, *What Computers Still Can't Do* (Cambridge: MIT Press, 1993); John R. Searle, "Minds, Brains, and Programs," *Behavioral and Brain Sciences* 3 (1980): 417–24.

46. Gemignani, "More on the Use of Computers by Professionals"; John M. Mulvey and Deborah Johnson, *The Ethics of Large-Scale Computer Decision Procedures in the Public Sector* (Princeton: School of Engineering and Applied Science, Department of Civil Engineering and Operations Research, 1992); Marshal S. Willick, "Professional Malpractice and the Unauthorized Practice of Professions: Some Legal and Ethical Aspects of the Use of Computers as Decision-Aids," *Rutgers Computer and Technology Law Journal* 12 (1986): 1–32.

47. Marcel C. LaFollette, *Stealing into Print: Fraud, Plagiarism, and Misconduct in Scientific Publishing* (Berkeley: University of California Press, 1992); Harry Redner, "Pathologies of Science," *Social Epistemology* 1 (1987): 215–47; Warren Schmaus, "Fraud and the Norms of Science," *Science, Technology, and Human Values* 8 (1983): 12–22; Harriet Zuckerman, "Norms and Deviant Behavior in Science," *Science, Technology, and Human Values* 9 (1984): 7–13.

48. Blaise Cronin, *The Citation Process: The Role and Significance of Citations in Scientific Communication* (London: Taylor Graham, 1984); Kaplan, "The Norms of Citation Behaviour"; Nelson and Pollock, eds., *Communication among Scientists and Engineers*; Derek J. de Solla Price, "Networks of Scientific Papers," *Science* 149 (1965): 510–15.

7. Finding the Space in the Text and the Text in Space

1. Henri Lefebvre, *The Production of Space*, trans. David Nicholson-Smith (New York: Basil Blackwell, 1991), p. 48. Subsequent page references will be given in the text in parentheses.

2. Here I refer, of course, to Thomas Nagel, *The View from Nowhere* (New York: Oxford University Press, 1986). With respect to issues in geography, his work is used in J. Nicholas Entrikin, *The Betweenness of Place* (Basingstoke: Macmillan, 1991).

3. I have found Ted Cohen's "Metaphor and the Cultivation of Intimacy," in Sheldon Sacks, ed., *On Metaphor* (Chicago: University of Chicago Press, 1979), pp. 1–10, a useful and provocative expression of this view.

4. Here see, for example, Thomas L. Haskell, "Professionalism versus Capitalism: R. H. Tawney, Emile Durkheim, and C. S. Pierce on the Disinterestedness of Professional Communities," in Thomas L. Haskell, ed., *The Authority of Experts: Studies in History and Theory* (Bloomington: Indiana University Press, 1984), pp. 180–225; Thomas L. Haskell, *The Emergence of Professional Social Science: The American Social Science Association and the Nineteenth-Century Crisis of Authority* (Urbana: University of Illinois Press, 1977); and Emile Durkheim, *Professional Ethics and Civic Morals* (Glencoe, Ill.: Free Press, 1958).

5. For mathematics, see David Bloor, "The Living Foundations of Mathematics," *Social Studies of Science* 17 (1987): 337–58; Philip J. Davis and Reuben Hersh, "Rhetoric and Mathematics," in John S. Nelson, Allan Megill, and Donald McCloskey, eds., *The Rhetoric of the Human Sciences: Language and Argument in Scholarship and Public Affairs* (Madison: University of Wisconsin Press, 1987), pp. 53–68; Eric Livingston, *The Ethnomethodological Foundations of Mathematics* (Boston: Routledge and Kegan Paul, 1986); and Derek L. Phillips, "Wittgenstein und die Soziologie der Mathematik," *Kölner Zeitschrift fur Soziologie und Sozialpsychologie* 6 (1975): 2–78. For physics there is a large literature, but see especially, Andrew Pickering, *Constructing Quarks: A Sociological History of Particle Physics* (Chicago: University of Chicago Press, 1986).

6. Karin D. Knorr-Cetina, *The Manufacture of Knowledge: An Essay on the Constructivist and Contextual Nature of Science* (Oxford: Pergamon, 1981).

7. W. V. O. Quine, *From a Logical Point of View* (Cambridge: Harvard University Press, 1953); Gilbert Ryle, *The Concept of Mind* (New York: Barnes and Noble, 1949); and Ludwig Wittgenstein, *Philosophical Investigations,* 3d ed. (New York: Macmillan, 1968).

8. Friedrich Nietzsche, "On Truth and Lie in an Extra-moral Sense," in Walter Kaufmann, ed., *The Portable Nietzsche* (New York: Viking Press, 1954), pp. 46–47; as quoted in Lefebvre, *The Production of Space,* p. 138.

9. See Thomas S. Kuhn, *The Structure of Scientific Revolutions,* 2d ed., enlarged (Chicago: University of Chicago Press, 1970); and especially Thomas S. Kuhn, "The Function of Measurement in Modern Physical Science," *Isis* 52 (1961): 162–76; and Thomas S. Kuhn, "Second Thoughts on Paradigms," in Frederick Suppe, ed., *The Structure of Scientific Theories,* 2d ed. (Urbana: University of Illinois Press, 1977), pp. 459–82.

10. For an interesting discussion of this, see Arthur C. Danto, "Narrative Sentences," *History and Theory* 2 (1962): 146–79; and "Mere Chronicle or History Proper," *Journal of Philosophy* 50 (1953): 173–82.

11. Hayden White, *Metahistory: The Historical Imagination in Nineteenth-Century Europe* (Baltimore: Johns Hopkins University Press, 1973).

12. The best summary is the often reprinted and revised Robert K. Merton, "The Normative Structure of Science," in *The Sociology of Science: Theoretical and Empirical Investigations,* ed. Norman W. Storer (Chicago: University of Chicago Press, 1973), pp. 267–78.

13. Michel Foucault, "What Is an Author?" in Josué Harari, ed., *Textual Strategies: Perspectives in Post-Structuralist Criticism* (Ithaca, N.Y.: Cornell University Press, 1977), pp. 141–60.

14. I find only four citations to the works of women, and only Julia Kristeva receives more than passing mention.

15. Diana Crane, *Invisible Colleges* (Chicago: University of Chicago Press, 1972).

16. See, for example, Henry G. Small and B. C. Griffith, "The Structure of Scientific Literatures I: Identifying and Graphing Specialties," *Science Studies* 4 (1974): 17–40; Richard D. Whitley, "The Formal Communication System of Science: A Study of the Organisation of British Social Science Journals," *Sociological Review* 16 (1970): 163–79; and Nan Lin, William D. Garvey, and Carnot E. Nelson, "A Study of the Communication Structure of Science," in Carnot E. Nelson and Donald K. Pollock, eds., *Communication among Scientists and Engineers* (Lexington, Mass.: D. C. Heath, 1970), pp. 23–60.

17. Henry G. Small and E. Sweeney, "Clustering the Science Citation Index Using Co-citations," *Scientometrics* 7 (1985): 391–409; Daniel Sullivan, D. Hywel White, and Edward J. Barboni, "Co-citation Analyses of Science: An Evaluation," *Social Studies of Science* 7 (1977): 223–40; Derek J. de Solla Price, "The Analysis of Square Matrices of Scientometric Transactions," *Scientometrics* 3 (1981): 55–63; Edward Nadel, "Multivariate Citation Analysis and the Changing Cognitive Organization in a Specialty of Physics," *Social Studies of Science* 10 (1980): 449–73; Timothy Lenoir, "Quantitative Foundations of the Sociology of Science: On Linking Blockmodeling with Co-citation Analysis," *Social Studies of Science* 9 (1979): 455–80.

18. In this group I include, for example, R. Shields, "Social Spatialization and the Built Environment: The West Edmonton Mall," *Environment and Planning D: Society and Space* 7 (1989): 147–64; Andrew Herod, "From Rag Trade to Real-Estate in New York Garment Center: Remaking the Labor Landscape in a Global City," *Urban Geography* 12 (1991): 324–38; C. Katz, "Sow What You Know: The Struggle for Social Reproduction in Rural Sudan," *Annals of the Association of American Geographers* 81 (1991): 488–514; and P. Routledge, "Putting Politics in Its Place: Baliapal, India, as a Terrain of Resistance," *Political Geography* 11 (1992): 588–611.

19. G. Burgel and M. G. Dezes, "An Interview with Henri Lefebvre," *Environment and Planning D: Society and Space* 5 (1987): 27–38; Edward W. Soja, "Henri Lefebvre, 1901–1991," *Environment and Planning D: Society and Space* 9 (1991): 257–59.

20. E. W. Soja, "Regions in Context: Spatiality, Periodicity, and the Historical Geography of the Regional Question," *Environment and Planning D: Society and Space* 3 (1985): 175–90; A. Pred, "Survey-14: The Locally Spoken Word and Local Struggles," *Environment and Planning D: Society and Space* 7 (1989): 211–33.

21. For example, S. Britton, "Tourism, Capital, and Place: Towards a Critical Geography of Tourism," *Environment and Planning D: Society and Space* 9 (1991): 451–78; A. Amin and K. Robins, "The Reemergence of Regional Economies: The Mythical Geography of Flexible Accumulation," *Environment and Planning D: Society and Space* 8 (1990): 7–34; N. Smith, "History and Philosophy of Geography: Real Wars, Theory Wars," *Progress in Human Geography* 16 (1992): 257–71; and Shields, "Social Spatialization and the Built Environment."

22. Jane C. Ginsburg, "French Copyright Law: A Comparative Overview," *Journal of the Copyright Society of the USA* 4 (1989): 269–85, is a useful summary. Also useful are Raymond Sarraute, "Current Theory on the Moral Right of Authors and Artists under French Law," *American Journal of Comparative Law* 16 (1968): 465–86; Rudolf Monta, "The Concept of 'Copyright' versus the 'Droit d'Auteur,'" *Southern California Law Review* 32 (1959): 177–86; and R. J. DaSilva, "'Droit Moral' and the Amoral Copyright: A Comparison of Artists' Rights in France and the U.S.," *Bulletin of the Copyright Society of the USA* 28 (1980): 1–58. Although now rather dated, Martin A. Roeder, "The Doctrine of Moral Rights: A Study in the Law of Artists, Authors, and Creators," *Harvard Law Review* 53 (1940): 554–78, is still worth looking at.

23. See Ernest Cushing Richardson, *Classification: Theoretical and Practical* (New York: H. W. Wilson, 1930); W. C. Berwick Sayers, *An Introduction to Library Classification*, 6th ed. (London: Grafton, 1943); in geography, Christopher Merrett, *Map Cataloguing and Classification: A Comparison of Approaches* (Sheffield: University of Sheffield, Postgraduate School of

Librarianship and Information Science, 1976); and for an account relating classification to the practice of science, Anne Brearley Piternick, "Traditional Interpretations of 'Authorship' and 'Responsibility' in the Description of Scientific and Technical Documents," *Cataloguing and Classification Quarterly* 5 (1985): 17–33.

Conclusion: Learning from the Place of the Work in the World

1. I recognize, of course, that several recent works have offered far more clearly articulated views of style in science; I have in mind A. C. Crombie, *Styles of Scientific Thinking in the European Tradition: The History of Argument and Explanation Especially in the Mathematical and Biomedical Sciences and Arts,* 3 vols. (London: Duckworth, 1994); Johan Galtung, "Structure, Culture, and Intellectual Style: An Essay Comparing Saxonic, Teutonic, Gallic, and Nipponic Approaches," *Social Science Information* 20 (1981): 817–56; and Ian Hacking, " 'Style' for Historians and Philosophers," *Studies in the History and Philosophy of Science* 23 (1992): 1–20.

Index

Michael R. Curry is an associate professor in the Department of Geography at the University of California, Los Angeles. He has held visiting appointments at the University of Edinburgh, Rutgers University, and Harvard University. He received his Ph.D. in geography from the University of Minnesota, and holds earlier degrees in geography, philosophy, and liberal arts.